D0319609

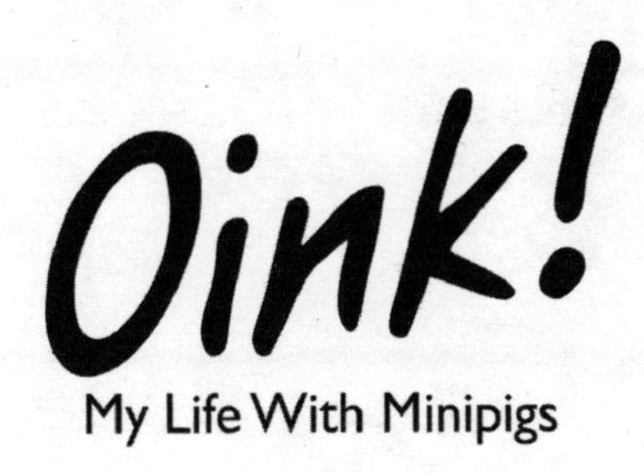
Oink!
My Life With Minipigs

MATT WHYMAN

Oink!

My Life With Minipigs

HODDER &
STOUGHTON

First published in Great Britain in 2011 by
Hodder & Stoughton
An Hachette UK company

1

A CIP catalogue record for this title is
available from the British Library.

ISBN 978 1 444 71144 8

Typeset in Celeste by Palimpsest Book Production Limited,
Falkirk, Stirlingshire

Printed and bound in the UK by Clays Ltd, St Ives plc

Hodder & Stoughton policy is to use papers that are natural, renewable and recyclable products and made from wood grown in sustainable forests. The logging and manufacturing processes are expected to conform to the environmental regulations of the country of origin.

Hodder & Stoughton Ltd
338 Euston Road
London NW1 3BH

www.hodder.co.uk

This book is dedicated to my wife, Emma.
Without her, none of this would've happened.

Contents

A Prologue to the Pigs 1

PART ONE

1. On Children and Animals 11
2. A Real Man's Best Friend 16
3. An Escalation of Pets 22
4. Size Matters 26
5. A Crash Course in Pig Keeping 34
6. Pride and Preparation 44
7. Enter the Minipigs 54
8. In-House Training 68
9. How Much Trouble Can They Be? 78
10. More Than a Match 89
11. An Embarrassment to the Family 98

PART TWO

12. A Drunk with a Digger 115
13. This Little Piggy Went to Market 125

14. Under Inspection 135
15. On Loss 145
16. Muscle and Sweat 155
17. The Jail Birds 164
18. A Bump in the Night 177
19. Cut to the Chase 191
20. Housebound 205
21. On Space 218
22. In the Wilderness 233

PART THREE

23. Tools for the Task 249
24. A Man about a Minipig 268
25. Short Straws 284
26. Nature and Gravity 299
27. Third Time Lucky 311
28. One of Our Minipigs is Missing 326

An Epilogue to the Pigs 347
Postscript 353
Acknowledgements 355
Author's Note 357

Always remember,
a cat looks down on man, a dog looks up to man,
but a pig will look man right in the eye and see his equal.

Winston Churchill

A Prologue to the Pigs

It's the moment every family-pet owner dreads. The kids come skipping home from school to find you waiting for them with a solemn face. They look around, wondering what's missing, and then you have to break the news. In my case, our beloved cat, Misty, had been struck down on a country lane outside our house.

'Guys, there's something I have to tell you . . .'

With four children in tears, the only comfort I could offer was the fact that Misty had suffered no pain. This I had on good authority, because the driver responsible for running her over turned out to be a vet. When the accident occurred, he'd been racing to attend a 'cowdown', whatever that meant. On handing over the corpse of our poor cat, considerately enshrouded in a towel, he assured me with conviction that it had been over quickly. I'd done my best to tell him these things happen. I just wished it hadn't occurred as I was rushing out for a dentist's appointment. So, I'd placed the bundle on top of the log shelter round the back of the house, with a view to dealing with it later.

Lying in the chair, while being lectured about flossing, I decided that I would make arrangements for a back garden burial. The last time the Grim Reaper of Small Animals knocked upon our door, in the form of a fox that savaged two of my three chickens, I had placed the remains in a feed bag and dumped them in the woods. The kids were horrified when they found out. In their view I had basically disposed of two members of the family by feeding them to predators. Things would be different this time, I thought to myself, beginning with a freshly prepared grave. This way, everything would be ready for us to do the decent thing. So, on my return, I chose a spot at the end of the garden where I had kept the hens, and began to dig. It wasn't the best place, I realised, on struggling to get through the clay and the roots from a towering oak beside the lane. We were just coming through a hard winter as well, and frost still coated the ground. After half an hour, having managed to create a small depression, I decided it was just about fit for purpose. With the kids' return imminent, all I had to do was rehearse what I would say. By the time the back door swung open, I was ready to tell them the last thing they wanted to hear.

Minutes after I'd ruined their day, my children filed out into the garden as directed. Being young teenage girls, Lou and May were free and easy with their tears. At six, Honey hadn't really digested what it all meant, while Frank, aged four, was simply happy to be standing beside the spade. I approached them sombrely with the bundle in my arms, but something didn't feel right. What I held so reverently no longer felt floppy. I had left it so long that rigor mortis was in full effect, I realised. This wouldn't have been a problem. It's just the hole I had managed to dig suddenly didn't look that accommodating for a stiffened cat with outstretched legs. Proposing that we delay the service until the corpse had softened up was out of the question. I looked at my children in turn, swallowed uncomfortably, and then placed poor Misty to rest as best I could.

Later, when my wife, Emma, returned home from work, I shared our sad news. For a tall blonde dressed for the boardroom, one who never took kindly to being described as Amazonian, she looked unusually close to tears. At the grave, we found that one of the girls had planted a cross made from decorated ice cream sticks. I thought it was touching but Emma was more concerned by the bigger picture.

'What's with the big mound of earth?' she asked, stepping around the plot.

'It's because of the problem with the paws,' I said. By the way she had quickly recovered her composure, I realised I would have to explain myself a little more. 'The cat is in there diagonally, alright? Upside down and diagonally. She wouldn't fit, and I didn't want to upset the children even more. So I told them it was how cats were buried traditionally. In Nepal.'

For a moment Emma looked like she was going to kick-start a domestic. Then, watching her eyes finally well up over our loss, I hoped a hug would afford me a reprieve. Deep down she knew I would always try to do the right thing for the family. Even if that meant exhuming the body under moonlight and starting all over again.

That Misty had left behind a brother was of little consolation. Miso was identical in appearance to his sister. Both black with white socks, they only differed in their personality. In short, Misty had been outgoing and fun to be around while Miso wasn't.

When it came to avoiding human contact, Miso was a master. In this house, at any rate. We suspected that he chose to work another family nearby, who clearly indulged him with treats. He also had a penchant for hunting wild rabbits, dismembering them, and leaving body parts outside the front door.

Within the bunny community, I imagine our surviving cat had a reputation as some kind of serial ripper. One summer, Miso was so prolific that I took to leaving a spade outside the door for the sake of convenience. Every time he delivered a severed head, hind leg, bones or entrails, I'd scoop it all up and fling it into the hornbeam hedge on the other side of the lane. This was a practice I had to stop come autumn, however, when all the leaves dropped to reveal the rabbits' remains. For a long time afterwards, I worried that passers-by would think we had strung up some kind of ghastly pagan offering to ensure that spring would come around again.

Showing up only at mealtimes hardly endeared Miso to me, and his indifference to the death of his sister failed to change my view. Our second born, May, was the only one who tried to identify with him. Nobody dared to point out that the cat clearly hated her as much as any of us. A sensitive child, she could see something in Miso that was invisible to everyone else. Whenever any kind of communication took place, the cat evidently saw her coming. May couldn't bear to see Miso's saucer empty, even if he'd just cleared it. Once, she found him lapping from a glass on her bedside table. It led her to believe that this was a cry for help. In May's view, the cat could die from dehydration unless she placed cups brimming with fresh water all around the house. Including random steps on the staircase. Miso went on to learn very quickly that what he really needed was space from me every time I kicked a cup over. Apart from the cursing, I didn't feel it was appropriate to spell out what I really thought about our one surviving feline. Such was the grief shared by my wife and kids over Misty's death, it would've just made things worse. Nevertheless, we all knew it. The wrong cat had been run over. Where Misty had brought joy into our home, Miso just brought us grief. I could only think he served to remind Emma that something was still missing from our lives.

Not for a moment did I consider what shape this would take.

I like order in my life, I'd be happy to settle for just that. With four children, peace and quiet are out of the question. In my book, we count as a big family. A *complete* family. As for my wife, she tends to look at what space is left and then fills it.

'What is the difference,' Emma asked me some days after Misty's burial, 'between a gerbil and a guinea pig?'

'I can tell you what they have in common,' I said. 'Neither have a place in this household.'

'But it would be great for the little ones,' she said. 'A pet that Frank and Honey could call their very own.'

'Hamsters and guinea pigs are hardly pets,' I replied, aware that the task of cleaning them out would invariably fall to me. 'They're always breaking. It's like buying cheap furniture. You always regret it in the end. If you're going to get another animal, at least make it count.'

'Like what?'

I turned to look out at the garden. In a coop at the back, across from where we had buried Misty, stood a forlorn-looking hen minus many tail feathers. I was still gutted about the fact that a fox had made off with Maggie's two sisters. What's more, we lived in a village where almost every household kept at least a couple of chickens. An outsider might be struck by the fact that all of them appeared to be from the same breed, and they would be right. The fact is these tawny-coloured utility hens had all come from one source. Namely the free-range poultry farm on the other side of the woods. The one with the hole in the fencing that the farmer never got round to repairing. It wasn't unusual to see the odd chicken roaming freely around the churchyard or even roosting upon a road sign. Whenever one wandered into a garden, often the owner would quietly claim it for the pot or their own poultry stock. On being offered three such runaways by an old boy down the lane with a flock too big for him to manage, I'd found it hard to refuse. Like the dog, who provided the family with security, three chickens served a purpose. They might not deter a burglar but they could offer us eggs. After everything she'd been through, however, Maggie couldn't actually manage that any more.

'How about something to deter a fox?' I suggested idly. 'Like a grizzly bear, a pond with piranhas in it, or maybe some kind of pig?'

'*Right*. A pig. In a domestic back garden.' Emma's gaze tightened. Just enough for me to know this was no time to be making stupid suggestions. 'If you want to keep foxes away all you have to do is urinate around the perimeter,' she told me. 'Apparently they pick up on the testosterone and run.'

'Do you really think that could work?' I ask.

Emma considered me for a moment.

'Probably not.'

Looking back, I blame Maggie the chicken. Had I not seen her looking so vulnerable and lonely, it would never have entered my head. All I know is that suggesting our surviving hen might like some company beyond a buried cat must have seeded the need for a little online research. How can I be sure? Because the next day I received an email from my wife at work. The image was bizarre: two pigs perched upon a twig. My first thought was that it couldn't be real. Then I read her message underneath, and a chill passed right through me:

soooooo cute!!!

Introducing Roxi. Note the ladylike lashes and shovel for a snout.

The mighty Butch.

Sesi the wolf hound. A surrogate mum to the minipigs.

PART ONE

Replica minipigs (not quite to scale).

Emma, just before heading out to a gala work do, bribing Roxi to be good.

1

On Children and Animals

Ever since she was a little girl, Emma always wanted a big family. She gave me plenty of warning. We were at primary school together, after all, though we didn't start dating until our twenties. By the time we got married, I knew she had plans to go beyond the national average when it came to having kids.

I took the view that three girls were quite enough. Emma countered that we needed a boy, and on that basis pushed it to four. I don't like to think what would've happened had we produced another child ready-made for pink hand-me-downs. Privately, my joy at the birth of a son was also fuelled by a massive sense of relief. At last, my job was done.

I also knew just who I needed to turn to in order to retire on a permanent basis.

Our doctor divided opinion. Some considered him to be a medical pioneer. Others marked him down as a psychotic self-harmer. When I learned that he had been the first man to perform a vasectomy on himself, I just crossed my legs and expressed disbelief. Why anyone would want to go there was beyond me. I figured it must've taken a heroically steady pair of hands or a huge intake of drugs. Either way, he had gone on to become evangelical in providing a walk in/hobble out service from the local surgery. Had anyone conducted a straw poll in the neighbourhood, I doubt they would've found a father of two point four children capable of producing any more. I had only escaped the doctor's clutches because Emma went to great lengths to shield me from him. For quite a while, I wasn't allowed to be ill. If I'd needed antibiotics, I reckon Emma would've faked the symptoms herself in order to get me the treatment. As far as she was concerned, I was not going to be another statistic in the doctor's fertility cleansing programme. But as soon as he was assigned to our son's first antenatal home visit, something Emma hadn't foreseen, I think even she knew the game was up.

Within weeks of Frank's arrival, I prepared to be put out to pasture. Leaving Emma in the waiting room, unusually subdued as she leafed through a magazine recipe for plums past their prime, I took steps to do the right thing. Our doctor needed Emma's written consent for this, which she provided without a word. Even so, before the swelling had subsided, it became clear to me that, in her view, such a simple procedure would have far-reaching consequences. She didn't make me feel guilty. Far from it. I received tender, loving care and cups of tea while I recovered. I just knew from the look in her eyes that it was a loss she suffered more than me.

Having known Emma since I'd been in shorts, I knew that her need for a sizeable brood stemmed from her experience of growing up. Her childhood home was not a happy one, and much of this was down to her father. A closed individual with a difficult upbringing, he was a man who found it hard to connect with those who looked to him for love. His wife and two daughters saw the good in him, but he just did not know how to show it unreservedly. Instead, he escaped into a solitary drinking habit that would come to overshadow his life. Emma's mother did her best to cover for him but their relationship took its toll on her health. Her slide into a fog of medication, along with periods of absence as work became her means of escape, left Emma and her younger sister to spend their formative years with front door keys around their necks.

I remember being envious of her independence. Emma could pretty much do whatever she pleased. In my eyes, unaware of the bigger picture, she was the true definition of a free spirit. It was only later that I realised she viewed my kind of background as something she would like to have experienced for herself.

Emma's most vivid childhood memory is of screaming in the street, aged five, because she had woken up one morning to find her parents were nowhere to be found. I can barely recall a moment when I was home alone. My mother worked for the NHS as a part-time physiotherapist, fitting her hours into a school day. It meant she was always there for my little brother, sister and me, while my dad was a BBC engineer and the kind of green-belt commuter who would come home to find supper on the table. Naturally, I didn't fully appreciate the stability at the time. Nor did I seize the opportunity to use it as a springboard into the world. I was a bit of a worrier when it came to taking on new challenges, while Emma was forthright and unafraid to stand up for herself. In a way, we were drawn together by what each of us lacked in ourselves. It had worked well over the years, and despite my wife's instinct to keep building the family she'd missed out on as a girl, she did agree finally that going to five was, well, *insane*.

'I can't help how I feel,' she reasoned, 'but I'm sure I can channel it into something other than more babies.'

In short, with the offspring ticked off the list it was time to take on some animals. There was only one obstacle to this newfound need. As we were crammed into a three-bedroom terrace in London's East End, one of which I used as an office, we had no room for anything other than ourselves. For a time we kept a couple of goldfish. Unfortunately, we rarely saw them. Despite my best efforts, the walls of the tank became increasingly caked in a stubborn film of algae. Every now and then, one would swim close enough to the glass for us to be reminded of their existence. Eventually, they came up for air and never went down again.

In some ways the fate of the fish came close to mirroring our financial situation. Several years earlier Emma had given up a good career to return to full-time motherhood. I kept the roof over our heads and put food on the table by writing books for children and taking what freelance journalism I could find. But with the increase in the number of mouths to feed, I began to struggle. We were OK: happy, but totally cash-strapped. So, when Emma was offered the chance to reignite her career on a part-time basis, the decision was largely driven by a bid to avoid living out of skips. That the job was outside London proved another draw. Here was our chance to raise our children in a rural county while not being so far away from the capital that they'd be freaked out by things like neon signs and elevators. One final thing persuaded me it was time to move on. I had discovered that the youths who frequently parked their car outside our house and opened up silver foil wraps on their laps weren't, as I thought, eating sandwiches. As Emma informed me when I made the observation, in front of her mother and toddler group, they were in fact dealing crack cocaine. With her job offer on the table, it didn't take long for us to agree that here was our chance to downsize to the countryside.

'If we sell this house we can just about find something with a garden,' suggested Emma. 'The kids could even keep a rabbit outside.'

I wasn't resistant to the idea at all. As I worked for myself, it meant I could juggle my hours between writing and being a house husband. Privately, I knew we could do a bit better than a bunny. From my childhood experience of looking after one, I just remembered rabbits to be high on maintenance and low on reward. By the time we left the city behind, I had set my sights on something more robust.

2

A Real Man's Best Friend

A dog made perfect sense. This became quite clear to us within a week of moving. We had fallen for a place on the edge of woods in a ragged region of West Sussex. It was on the crest of a lane leading out from a village, with a few houses peppered sporadically on each side. The property was badly run down with a scrappy, overgrown garden, all of which made it affordable. In fact, the whole purchasing procedure was fairly straightforward on account of the owner being deceased. Unoccupied for eighteen months, it was one of those places where the interior space and location ticked the right boxes, while the sense of abandonment was something we just had to see beyond. Having no curtains didn't help. Nor did the fact that we were miles from any street light. If Emma and I were going to sleep soundly at night, we needed something to protect us from what we regarded as major threats to our safety: the darkness and the silence.

After ten tranquil years of inner-city living, every sundown over the trees here felt like a remake of *The Blair Witch Project*, and so it was agreed that a canine companion would serve a useful purpose. Emma and the children campaigned for a labradoodle. I didn't see how something that camp-looking was going to grant us peace of mind. Knowing I would be the one to walk it, I argued for a real man's best friend. I couldn't face taking out a curly-haired hybrid; that screamed *wrong.* I wanted something noble, upright and strong. I just wasn't clued up enough about dogs to know what breed would fit the bill.

The answer came that summer, during a camping holiday in France. We were on a beach, fooling around with sandcastles, when a white wolf emerged from the pine trees that pinned the shore to the sea.

'Get back to the tent and zip yourselves in!' I ordered the children, hefting a plastic spade in my hands as if it would protect me from rabies. 'No sudden movements. Just take it very easy!'

For a moment, I watched in awe as the creature loped down the beach. As soon as I spotted the collar around its neck, I realised that I was in fact looking at a domestic dog of some description. I had just never seen anything so formidable in my life. With its square-cut muzzle, arching ears, muscular haunches and sloping tail, it really did look like something from a fairy tale. The dog's presence was certainly enough to turn heads among the other holidaymakers. Admittedly, I didn't see anyone else usher their family to safety. On turning to check mine were out of harm's way, I realised they had merely gravitated towards their mother for protection.

'Stand down, soldier,' said Emma. 'It obviously belongs to someone.'

Sure enough, the creature in question had just dropped down beside a woman on a beach towel with a mobile phone in hand. Obediently, it sat there looking out across the water as if on guard for some sea-borne strike. Right then, I knew that I had found my kind of dog. I just had to find out the name of the breed. The trouble was the owner in question happened to be young, beautiful, oiled and topless. I couldn't simply go sauntering across the sand without looking like I was scoring on her. With pleading eyes, I turned to Emma.

'Oh, go on,' I said. 'She won't feel threatened if you ask.'

'I hardly think you present a threat,' she replied. 'You're the one who wants to find out. Don't let me stop you.'

I sighed to myself, wishing I had come to the beach wearing more than a pair of over-sized, regulation Brit-abroad shorts.

'Very well,' I said, sucking in my belly a little bit. 'Just don't blame me if she falls for my charms.'

The woman in half a bikini became aware of my presence at the same time as her dog. Only one of them growled. I could've walked on, but such was my determination that I ignored the loss of feeling in my legs and stopped before them both. I drew breath to introduce myself, and my innocent reason for coming across. What prevented me was the fact that she still had the mobile phone pressed to her ear. Backing out was not an option. That would've been weird. All I could do was stand there and wait for her to finish the conversation. I had hoped she would take just a second or so to close the call. When a minute passed, it felt more like a millennium. While she focused on the sand and did her level best to ignore my presence, the dog simply stared at me with baleful eyes. As for me, I just did not know where to look. Every time I shifted my gaze, somehow it felt like I was trying to sneak a crafty glimpse of her breasts. Even admiring the clear blue sky felt like an invasion of her privacy. Eventually, to my great relief, she snapped her phone shut.

'Hi!' I said brightly, while dying inside. 'I was just admiring your dog. I'm really sorry to trouble you, but could you tell me about it?'

The woman said nothing for a beat. Then, much to my discomfort, she crossed her arms to cover her chest. I even found myself doing the same thing; clasping my ribs as she did. She looked unsettled, but thankfully not alarmed. Maybe that was because the dog curled back its slack black lips and showed me its fangs.

'She's my *bebe*,' the woman said eventually, in a heavy French accent. 'A *she* dog.'

'Is it? Right. OK. Thanks! Sorry to bother you.' I began to retreat, hating myself for failing, having come so far. It wasn't exactly the depth of information I'd been holding out for, after all. Only as I turned, however, did she appear to take pity on me.

'She's a Canadian Shepherd, you know? Like a German Shepherd, but not so the same.'

I came around full circle to find the woman had risen from her towel. She was still covering herself, which encouraged me to lock my attention on the dog. It was no longer growling now, but watching its owner move towards the water. I raised my eyebrows at the beast, as if to seek permission to follow. The dog climbed to its feet, clearly unwilling to leave me alone in her company.

'Does she bite?' I asked, on catching up with her.

'It depends.'

Warily, the dog glanced across at me, and then left me behind to pad into the shallows.

I watched the woman sink into the water. Half of me wondered whether she was inviting me to follow her. The other half that wasn't living in a total dream world knew for sure she was trying to create some distance. Once submerged up to the neck, she turned to face me once more. I didn't speak much French, but now I could read her body language fluently. I swapped a look with her dog. Even if I had been the sort of person with designs on her, it was clear I wouldn't get so much as a toe into the water without having my throat ripped out.

'I really just want to know about this fine animal of yours,' I assured her, spreading my palms. 'Really and truly. I'm here with my wife.'

I gestured behind me, and then glanced around. I had assumed Emma would be watching me intently. Instead, she had returned to building the sandcastle with the kids. Evidently she wasn't concerned about my potential to seduce. I turned back to the woman in the water. Judging by her wry smile, it seemed she had drawn the same conclusion.

'OK,' she said. 'Ask me anything.'

Over the course of the next ten minutes, despite the language barrier, I learned a great deal about Canadian Shepherds. By the time we said goodbye, I was convinced that a dog like this was what we needed to protect our family.

'That seemed to go well,' said Emma, when I joined her on the groundsheet. 'What's her name?'

'She's a bitch,' I said. 'That's all I know.'

Emma looked at me side on.

'What did she do? Make fun of your shorts?'

'Oh, the woman didn't say much about herself, probably for personal security reasons. But I can tell you I will be going on a mission to find us a Canadian Shepherd. Apparently they're great with kids, and with none of the aggression of German Shepherds.'

'So why was that one showing you its teeth?'

I reached for the sunblock.

'It was doing what comes naturally. A dog like that is fiercely loyal. When the children are a bit older they could take it into the woods for a walk and we wouldn't have to worry about a thing.'

As I spoke, the woman and the Canadian Shepherd returned to their spot on the beach. I had fallen in love. Emma could see it in my eyes.

'You really want a dog like that, don't you?'

'It's fate,' I said. 'How do you feel about it?'

'It's some responsibility,' she replied. 'Are you sure?'

'We have so many children that we can't even fit in a normal family estate car,' I reminded her. 'How much extra work can it be?'

3

An Escalation of Pets

Back home, sourcing a puppy became my pet project. While Emma prepared to return to work, I spent my free time scanning classified adverts in pedigree dog magazines. Once I found what I'd been looking for, and sealed the deal over the phone, it was just a question of counting down the days before we could collect our very own White Canadian Shepherd puppy. I even settled on an appropriate name before I'd set eyes on her. With such a striking white coat, I figured something to do with snow would be fitting. Feeling artistic, I turned to the internet for an Eskimo term. Sesi struck me as fitting. The name wasn't hard to find, even though I had assumed it would take some canny research. As it turned out, I picked it up from a website called *Eskimo Names for your White Canadian Shepherd.*

Sure enough, just as the woman on the beach had promised, our new arrival was a darling. For the first few weeks at least. When our seal-like pup started growing sinew and teeth, and then rounding up the children, I figured she might be a handful. As Sesi became more like the wolf I had witnessed prowling the sands, only less submissive to her master, I began to worry that I might've bitten off more than I could chew. In the space of three months, we stopped being afraid of the dark and switched our fears to the dog.

One evening, all but barricaded upstairs with the wife and kids, I realised something had to be done. I took my responsibilities seriously, of course. Despite the half jokes, I had no intention of having Sesi put down or rehoused. I was responsible for a difficult dog. It was down to me alone to do whatever it took to ensure she found her place within the family. What it took involved much of my life savings. At a visit to the local dog training school I learned that what Sesi needed was a month-long residential reprogramming. The trainer, a man who kept a Rottweiler named Satan as a kind of calling card, assured me he could bring out the best in her. Figuring my family also needed the break, I wrote the cheque and swore to myself I would never take on another animal for the rest of my days. Four weeks later, Sesi returned to me as a different dog. According to a parting comment from the trainer, one I swallowed bitterly, she now had the disposition of a labradoodle.

Despite Sesi's newfound obedience, the experience left me in the doghouse with my family. I'd had my chance to choose a pet. Now it was their turn.

First came the kittens. We'd had some field mice in the loft, so it did make sense. I put forward just two reservations. Firstly, we now lived at the top of a quiet country lane. It was the kind that could see no vehicles whatsoever for an hour or more. Then, when a car did appear, it would hurtle over the crest as if completing the final leg of the World Rally Championship. In my view, a cat caught in the headlights would stand no chance. My second reason for being less than keen on cats took the shape of the dog.

On Sesi's return from canine rehab, the trainer offered me some advice. Such was her size and spirit, he suggested, it would be safest for everyone if she had some space of her own. This wasn't down to a fear that she would turn on the kids. It was evident to one and all that she loved them to pieces. What concerned the trainer was the risk of her trampling them with affection. So, having fitted child gates on the doors into the kitchen and my office, Sesi now occupied a kind of boot room/mud airlock. The arrangement worked wonders. The dog was still at the heart of the family. She just couldn't dominate it. Nevertheless, I had no doubt that one glimpse of a feline mincing through the house would send her into a frenzy. Frankly, I just didn't need the grief.

'If you get cats, it'll be your responsibility,' I told Emma. 'Should Sesi tear them to pieces, I am not liable.'

'I'll have my lawyers contact yours,' she replied, before making the call to a friend whose pedigree puss had been knocked up by a stray.

In retrospect, I should have known that my wife would not stop there. Soon after the kittens arrived, without due warning or negotiation, Emma upped her game with the rabbits she had promised. One for each child, to be exact. By now, in terms of pets, she had come close to breaking me. I even helped in setting up the hutches in the yard and their runs out in the garden. The kids loved all four bunnies equally. It was just that the kind of love they showed didn't extend to feeding them regularly, cleaning them out, or closing them in securely so Miso couldn't slaughter them. Quietly, I assumed a kind of support role to ensure they didn't die.

Before long I began to hold out hope that the rabbits might perish prematurely. Unlike the dog and the chickens, they offered nothing in return. What with their daily demands, including ferrying them to and from the garden for exercise, it felt more like I was caring for invalids. Emma would argue that the bunnies made the kids happy. As I saw things, their hutches took up much of the yard and the runs left us with little space to sit outside.

'Everywhere I look there are cages,' I complained. 'It's like *Watership Down* meets Guantánamo Bay out there.'

'We'd have more room without the chicken fencing,' Emma suggested. 'Why don't you get rid of it?'

'Because we still have a chicken,' I reminded her. 'Maggie might be on her own, but it's our duty to give her a good life.'

I did think about getting a couple more hens to keep Maggie company and also provide more eggs. What stopped me was the threat of a return visit from the fox. More importantly, I knew that it would prompt the animal equivalent of an arms race. If I could take on more chickens, Emma would regard it as justification for something else that looked adorable but was essentially incontinent. I only had to consider that worst-case scenario to pass on the prospect of additional poultry.

Besides, even Emma could see we were operating at capacity here.

Then the speeding vet reduced the feline contingent by one. It was undoubtedly a sad loss. For my wife, it was also an opportunity. With our number down, here was a chance for her to end the pet standoff once and for all. The stealth moves she went on to make in a bid to swell the animal count would have far-reaching consequences. Not just for me but for every member of the family. Even Emma herself was unprepared for what would become animals of mass distraction.

4

Size Matters

Shortly after Emma emailed me the image of two pigs on a twig, the offensive campaign unrolled.

'Kids,' she began on her return from work, showing them the picture she had printed off. 'How would you feel about getting a *mini*pig? They've been especially bred to be as teeny tiny as possible.'

'Is this wise?' I asked as the children cooed collectively. 'Shouldn't you wait a while and see if they do a nano version?'

'Any smaller and we'd risk losing it!' she declared happily. With her enthusiasm in full effect, Emma made way for the children as they gathered around the picture. 'This is the perfect size for us.'

I refused to be railroaded. Previously, I'd just let my wife get on with it, but not this time.

'Emma,' I said patiently. 'We've already made the mistake of taking on animals with the potential to eat each other. Just look at the efforts we have to make to keep the dog from the cat and the cat from the rabbits. Inviting a little pig into the family wouldn't just be adding to the pet count. You'd be increasing our personal food chain.'

'We've never had an incident,' she countered.

'At a cost.' I gestured at the dog in the boot room. She was watching us through the bars of the child gate. 'Sometimes it doesn't feel like we live in a house. It's more like a series of holding pens.'

'But it works.'

'Yes, but we can't even let the little ones out in the yard if Sesi is out there. She's just too boisterous. What's it going to be like with a little pig running around?'

'It would be *fun*,' stressed Emma, as if I might not be familiar with the word.

Sometimes you just know when the deal is done. Every reservation I could muster was met by a pre-planned response. The proposed pet was highly sociable, good with hens and other animals but would deter Mister Fox. The daily upkeep would be conducted by Emma and the children. They had learned their lesson with the rabbits, they assured me. This time, things would be different. I would not have to be involved in any way. Pigs were also intelligent, so I learned, even the small model. They clocked in as one of the smartest species on the planet after humans, chimps and dolphins. According to the kids, piqued by my lack of enthusiasm, that made them sharper tools than me.

'Studies even show that all breeds of pig can dominate at videogames with joysticks,' Emma added, in a brazen aim for my weak spot. 'Just think about it. At last you can have some company on the Playstation.'

They had everything covered. Everything but the price tag. This detail Emma slipped in at the last moment, along with the fact that our name was on the list. Not just for one. For *two*.

'Out of the question!' I declared, grasping for a reason. 'There's only room for one pig in this household, and . . . erm, that's me!'

In the silence that followed, as my family left me feeling

like an old dog before it's put to sleep, I swore I could hear the patter of tiny trotters preparing to run riot through my life.

Exactly what is a minipig? I'd never heard of such a thing. Could it be a giant con, I asked myself? Searching online was quite an eye-opener. Once I'd got beyond the blog entries about these pint-sized porkers, usually marked by, *'OMFG!! I so want one!!'*, I found a biomedical website that immediately cut through the cuteness. Minipigs were real, I learned, only they hadn't been invented to make women and children go weak at the knees. In 1960s Germany a breeding programme had been undertaken to create a pig for one purpose only. Part pot-bellied Vietnamese, known for its fertility, and part Minnesota, recognised as being one of the most laid-back varieties on earth, the resulting minipig came into existence to meet the demands of science.

'These pigs,' I told Emma. 'They're basically lab rats.'

I showed her the website. In silence, she clicked through from the brief history I had read, only to be confronted by an image of a tiny pig with electrodes attached to its head. I couldn't say for what purpose because Emma shut down the browser in a blink.

'All the more reason we should give two a home,' she said. 'I've done my own research too, you know. This breeder has been in the business for years. She raises her own special brand of minipigs, but not so they can have shampoo squirted into their eyes. They're pets, pure and simple, and I'm not changing my mind on this. Pinky and Perky are going to be part of our family.'

I watched her lips shape the two names. I didn't need her to repeat what I thought I had heard.

'Oh, please. Is that the best you can do?'

'The little ones suggested it,' replied Emma. 'Are you going to overrule them?'

'In the interest of good taste, I have no choice.'

'From the man who named the chickens Maggie, Marge and . . . remind me of the last one?

'That had an original theme,' I countered. 'A tribute to *The Simpsons* and *Futurama*.'

'But do you think Bender was the most appropriate name you could've chosen? The kids went into school and told their teachers, you know.'

'Bender was a lovely chicken,' I said. 'So was Marge.' I paused there, and glanced out of the window. Ever since the fox attack, the sight of our one surviving chicken at the back of the garden never failed to make me feel bad. I couldn't genuinely hold Maggie responsible for the impending arrival of two minipigs. I just wasn't sure I could live with the crushingly clichéd names my wife was threatening for them. 'How about we keep the theme going?' I suggest. 'Bart is cool. So is Mister Burns. Or, how about Leela?'

'How about learning to live with Pinky and Perky?'

Emma folded her arms. In my mind, I foresaw a day when one escaped, probably the latter, and I would have to roam the village calling its name out loud.

Unlike a budgie or a hamster, you can't just pop into the pet store and pick up a minipig. Nor is there anywhere like the number of breeders in the country as there is for cats or dogs. As a result, you register interest, cry at the last of the savings you're asked to plunder, and then wait for a litter to arrive. It doesn't stop there, however. With a pig of any size comes paperwork. Not that I knew this beforehand. In fact, it only became apparent when I opened the post one morning over breakfast.

'What's this?' I showed Emma the form that had been sent to me. 'It's from the Department for the Environment, Food and Rural Affairs. That's DEFRA, isn't it? The Mad Cow people.'

Emma took the form, scanned it for a moment, and then appeared to remind herself of something.

'I forgot to tell you,' she said, and handed it back to me. 'You need to register as a farmer.'

'That's a good one,' I chuckled, and stopped when she showed no sign of joking. 'Really?'

'I called them the other day. By law, you have to notify DEFRA if you're going to keep swine. They told me you needed a registration number. It's very straightforward.'

'Emma, I'm not a farmer. I write children's books. If I wanted to get excited by tractors or shout at ramblers I'd have pursued a different career.' I stopped there to drink my tea as indignantly as possible. 'Anyway, why do I need to be the farmer? They're going to be *your* minipigs. *You* be the farmer.'

'I would,' she said, 'but you're at home every day. It just makes sense, in case they need to get in contact or anything.'

I studied the form. 'So what else does this qualify me to do? Blockade the lane with sheep carcasses and set light to them?'

'That was the French farmers,' Emma pointed out, before collecting a pen from beside the phone. 'And this just needs your signature.'

We didn't have to wait long before news came that Emma's minipiglets had been born. In fact, the breeder emailed us regular photographic updates on the progress of the litter.

When she sent us the first image, I took one look and wondered whether we had basically been sold shaved vermin.

'I'm not a complete fool,' I declared. 'You should ask for your deposit back.'

The shot in question showed a whole bunch of new arrivals feeding from a saucer of milk. Without meaning to sound distasteful, they were each the length of a sausage.

'They are minipiglets and make no mistake.' By her tone, I could tell that Emma did not want to believe anything else. At that moment, I really felt as if nothing would persuade her to rethink this upgrade to her pet plan. Maybe she could see the despair in my eyes, because her expression softened considerably.

'If it helps,' she offered, 'I'm prepared to rethink calling them Pinky and Perky. Actually, I've been wondering about the names you repeatedly suggested before each child was born.'

'The two you always overruled?' I brightened at this. Having done nothing but object to the whole minipig venture, this was the first instance where I felt a bit more positive. 'Butch,' I said, as if road testing how it sounded. I nodded to myself, smiling as I tried the two names together. 'Butch and Roxi.'

Over the next few weeks whenever new images arrived I would study them carefully and question their authenticity.

'Butch certainly looks small,' I said one time, 'but how do you know that bucket hasn't been placed in the foreground? Surely that's a normal sized piglet with a bucket positioned so it looks tiny.'

'Why would the breeder do that? Her website has loads of testimonies on it.'

'And another thing,' I persisted. 'If Butch and Roxi are from the same litter, how come they look so totally different?'

'They don't.'

'Butch is black with white trotters, so he's basically part cat,' I pointed out. 'And Roxi is the colour of a proper pig.'

I scrolled to a shot just to illustrate my point. The female had a few dark splodges on her, but otherwise her pinkness marked her out from her brother.

'And you became a minipig expert when?' asked Emma.

'Since I became a farmer,' I said. 'I think I have a responsibility to know where my flock have come from. If that's what you call a bunch of pigs.'

'It's a herd,' she told me. 'And the fact that they have different markings just shows their heritage. Minipigs are a combination of lots of different breeds nowadays. It means all kinds of aspects of the bloodline can come through in the same litter. Whatever it takes to create the perfect little snorter, that's fine by me.'

I accepted this without question. Why? Because I had saved the best evidence until last.

'If you'll just look at this shot with the breeder in it,' I said, moving onto the next image. 'You can't deny that the minipigs suddenly look much bigger. It just doesn't add up.'

Emma studied the picture. It featured a female figure in wellington boots sitting cross-legged on the floor of a barn. Although cropped at the neck, she was playing with piglets that most certainly could not fit inside teacups.

'They are quite big,' Emma agreed.

'Quite big? She's not short of a bacon sandwich there!'

I watched Emma take another look. At that moment, I felt a surge of righteous vindication. Then she tapped at the image on the screen and turned to me triumphantly.

'That's because it isn't the breeder. It's her daughter. She's five!'

With my eyebrows climbing, I faced the evidence once more.

'Is she tall for her age?' I asked, pretty much knowing there'd be no response.

5

A Crash Course in Pig Keeping

There is a sizeable age gap between our first pair of children and the next. It's enough to lead people to think we must each be onto our second marriage. We'd had Lou and then May within eighteen months of each other. We then left it for well over half a decade before repeating the process with Honey and Frank. The fact is that Emma always had a game plan to build a family of this size. It's just that until the second wave of infants arrived she also balanced motherhood with a career. The kind that involved wearing stilettos designed to kick through glass ceilings.

Having learned to stand on her own two feet from a very early age, Emma genuinely believed she could bring up children as well as break down Finance Directors until they gave her the budget she demanded for her department. Giving up the power suit and the business phone to have a third and then a fourth child wasn't easy for her. In fact it was a shock for both us to go back to changing nappies and wearing sick on our shoulders again, but we muddled through. Later, Emma's return to part-time employment was no surprise to me. The move was mostly for financial reasons, but quietly I knew my wife needed to prove to herself that she could manage.

What I didn't foresee, following our house move, was how

swiftly Emma's part-time post would snowball into a full-tilt commitment.

Time away from the kids left her torn, of course, and anxious to make it up to the family in any way she could. It also squeezed my opportunity to write for a living into the evenings. Still, the extra income kept our heads above water. It also obliged me to become a domestic god. A man at ease in a world of mothers. That was my ambition, but the truth is it wasn't easy being seen around the village during office hours. Despite being made to feel like I was unemployed or wearing an electronic tagging device, I managed to keep the last of our children alive through to school age. Once Honey and Frank were safe inside the classroom, my working days became my own again. At least until the bell sounded at three o'clock.

Where we failed miserably was in meeting the entertainment demands of the big and little league. This was what divided the family. We would take Honey and Frank to the playground, only for May and Lou to moan that they could be shopping with friends. So, we'd drive the older girls to the nearest town and feed two hours' worth of coins into the parking meter. Without fail, Honey and Frank would then misbehave so badly that we'd have vacated the space within ten minutes. Arguably the lowest moment came when Frank threw a tantrum on a busy pedestrian crossing. Unhappy that his family had bailed from Top Shop after he and Honey started chanting 'Mummy has a penis!' (which was metaphorically on the money with regard to her professional life), my son went on to throw himself upon the tarmac just as the lights turned from red to amber. Honestly, an airport protestor would've been impressed by Frank's ability to ignore the car horns and resist being moved on. It actually took the appearance of a traffic warden, and the quiet threat from me that he would go to prison, *forever,* before he finally came to his senses. In every way, ours was a family that creaked and strained like a tall ship on the high seas. We knew that we would get there. It was just a question of keeping the faith and staying on course.

Then came the minipigs.

A week before their arrival, watching my family's excitement mount, I was surprised to find myself thinking that things could be different. For once, here was something that brought everyone together. Everyone except me, of course, but then the prospect of feeding two more mouths didn't touch my heart so profoundly. Still, I felt increasingly uncomfortable casting one doubt after another when clearly Emma and the children were committed to welcoming Butch and Roxi into our world. I didn't want them to think of me as grumpy or obstructive. So, instead of pouring scorn on the stream of undeniably cute little piglet pictures that continued to arrive by email, I opted to offer my support. I would do so in the form of a gift, I decided. A gesture to prove that I was on board.

'Leave their accommodation to me,' I announced. 'I'll take care of it all. You don't have to worry about a thing.'

'That's sweet,' replied Emma. 'But what do you know about their needs?'

'Nothing. But I know a man who can help.'

The village where we live is just beyond the reach of a London commuter. It isn't so much the slog of the train journey. It's the hassle of getting to the nearest rural station and then finding a parking space. As a result, the majority of residents work locally. It means if we ever need anyone, from a painter and decorator to a rat catcher, we only have to visit the pub.

Which is directly where I headed with minipig housing in mind.

I have never been much of a practical type. Even self-assembly stuff defeats me. As I write, the desk I built several years ago is slowly falling apart in the manner of a clown's car. The undercarriage, if that is the correct term, drops away on one side whenever I shut the drawer too hard. Clearly I couldn't trust myself to build a shelter that would keep Butch and Roxi snug and dry. Despite my shortcomings, I knew an individual who could knock one up with his eyes closed.

Tom was born to fit the description of being 'good with his hands'. A landscape designer by trade, he could lay a patio in the time it would take me to stop idly surfing on eBay and start writing for a living. Standing at well over six foot, with the kind of weather-beaten features that mark out all men with proper, outdoor jobs, this was someone who knew how to handle a power tool. Tom also happened to keep pigs. Normal ones. He owned a field beside some woods along the lane. Originally a wasteland, he'd bought it at a knockdown price and then transformed it into a smallholder's paradise. First he built a stable block to house his wife's two horses, created paddocks for them to graze and canter, and then dedicated a strip beside the trees to his prize porkers. When Tom wasn't making lengths of timber bend to his will, he would be nursing a pint of ale at the Farrier's Arms. Sure enough, that's where I found him. As soon as I offered to buy him a drink, it was clear that I would be after something. In the past, Tom had helped to fence my chickens into an enclosure at the end of our garden and build a gate into it. My role had been to hold posts upright while he hammered them in before finding my skills redundant and making a lot of tea. This time, when I told him what I needed, the look on his face suggested I was asking him to build me a space ship.

'Minipigs?' He sipped his pint, mulling over my request. 'No such thing.'

With great patience, I explained that Emma had researched the subject extensively.

'I'm even a registered farmer now. DEFRA didn't dispute their existence.'

Tom sat back in his chair and bit on his thumbnail. I couldn't help noticing how his cuticles were encrusted with mud.

'What do you mean by mini?'

I spread my hands a little, only to give up with the estimation when Tom just smirked. 'Put it this way, the breeder has told Emma she can collect them in a cat basket.'

'Why do you want little pigs?' he asked. 'You won't get very big rashers off their backs.'

'I think Frazzles have cornered that market,' I said. 'Emma wants to keep them as pets. What I need is a pig ark for the garden, like yours, only smaller.'

'You're keeping pigs in the garden?' Using the sleeve of his fleece jacket, Tom wiped the froth from his top lip. 'They'll *trash* it. Pigs root on instinct. Their snouts are like shovels. They'll have the grass up in no time.'

All I could do at that moment was look at him helplessly.

'This isn't my idea,' I stressed. 'It's just time is running out now. They'll be here next weekend.'

Tom had just under half a pint left. He finished it with one drawn out gulp.

'You need a crash course in pig keeping,' he said, rising to his feet. 'There's still time to change your mind.'

*   *   *

To access his pigs' field, Tom led me through the horse paddock. Here, the grass had been neatly cut down to the stub. Mounds of manure littered the area, as it did in the stable we passed. It made me feel better about our imminent arrivals. At least there would be no mucking out to do on this scale

'Here we are.' On reaching the gate, Tom dropped his attention to the white trainers I was wearing. 'Let's hope they're washable, eh?'

Without reply, I followed Tom into the field. I could see that it was a bit muddy. I suppose I was so keen to spot the pigs themselves that I didn't think my feet might sink as deeply as they did.

'Bloody hell!' I muttered as my first step saw my footwear disappear into the soup. The suction was intense. If it hadn't been for Tom's steadying grip as I attempted to pull free, I'd have toppled backwards.

'This is the result of a winter with pigs,' he said, and helped me navigate a path to firmer ground. 'You get used to it.'

Finally, on standing upon the roots of a stranded-looking tree, I took in the full devastation surrounding me. Never before had I seen such a mess. It was worse than a building site. Every square inch of the field had been churned into a greasy, slate-coloured porridge. There were so many craters it looked like the site of a multiple meteor hit. Looking around, I realised that the animals responsible were nowhere to be seen.

'Where are the pigs?' I asked.

Without a word, Tom encouraged me to make the crossing to the woodland that bordered one side of the field. Abandoning all hope of keeping my trainers out of the mud, I followed dutifully in his bootsteps.

'Traditionally,' he said, halting at the treeline, 'farmers use swine to clear the undergrowth from land like this. Pigs love to forage, you see. It's what they're designed to do.'

I scanned the gloom between the trees. A slight breeze caused the leaves to shiver. All of a sudden, I felt a little uncomfortable.

'Why do I think I'm being watched?' I asked, and stepped back into the gloop.

'They can't see so well, but their hearing is razor sharp. You can be sure they'll be aware of our presence.' Stopping there, Tom pressed two fingers to his lips and issued a piercing whistle. 'Come out, my pretties. We have a visitor!'

In response, a solitary grunt rose up from the undergrowth. It sounded resonant and startlingly primal, like something from *Jurassic Park*. Specifically that moment before the Tyrannosaurus Rex made its presence known.

'Are we OK?' I asked to a snapping of twigs. 'I mean, safety-wise?'

Tom did not reply. He simply guided my attention to the frankly colossal shape that emerged towards the daylight.

'Hello, there!' he said cheerily.

'Good grief!' I felt my stomach tighten as the pig lumbered out towards us. I had never been so close to one before. From a distance, they just didn't look this big.

'It'll be cross when it realises we haven't got food,' said Tom. 'They know I often carry an apple in my pocket. Pigs love an old apple, but you have to be careful. If they're too ripe, and beginning to ferment, they can get drunk. And you don't want to be faced with a drunk pig.'

'I'm not sure I even want to be faced with a sober one, do I?'

'Just stay calm,' he said. 'Pigs can smell fear.'

I was losing any sense of reassurance by the second. As

several more hulking great swine trampled out through the undergrowth, the first one focused its attention on me. Hurriedly I patted my pockets.

'All I have is a chewy sweet,' I said, trying hard not to sound panicked. 'Offer yourself, Tom. *Do* something!'

By now, the pig was just a foot away. It was waist high to me, and clearly aware that I had something in my grasp. With its eyes locked onto mine, the beast began grinding its back teeth. Then, with a snort of what sounded like annoyance, it lowered its great head and shoved me in the crotch. Without waiting for permission, or wasting time with the wrapper, I tossed the sweet between its trotters. As the pig searched hungrily for the offering, I stepped around behind its owner.

'They'll eat everything,' Tom observed. 'Even human meat if that's what I chose to serve up.' At this, he glanced over his shoulder to find me looking up at him, aghast. 'I'm joking!' he grinned. 'Do I look like a murderer?'

'But they would,' I pressed him, 'if offered the chance?'

Tom shrugged, as if he wasn't bothered either way. 'Pigs are generally passive creatures. They're incredibly powerful, but they won't harm you. Not on purpose at any rate. Go on. Give him a stroke. He could use some affection. I'm taking them all to slaughter soon.'

Hesitantly, I reached out and patted the pig's flank. I had never realised they were so bristly. It felt as if I was petting a massive bottlebrush. I didn't like to think about the future it faced.

'Is this male or female?' I asked.

'You can see for yourself,' Tom suggested.

I peered down, mindful that two more pigs were now milling around me, and then stood tall with a start. 'OK, these are boys!'

'Their testicles can get very swollen,' said Tom.

'And muddy,' I added. 'I suppose there isn't much he can do about that, given how low slung they are.'

By now, all three pigs had realised we could offer them no more treats. I watched them sidle off, wincing as the lead pig ambled across a patch of brambles.

'They're a good size,' said Tom, though it took a moment before I realised he was referring to his herd. 'Soon, I'm going to have one very full chest freezer, plus a nice little profit from all my customers around the village.'

'How do you feel about that?' I asked. 'I'm not sure I could take them to the abattoir.'

Tom shrugged, still watching them. 'I don't give them names. That would be too much. Just food, water and fresh air. When they go, I know they've had a happy six months of life.'

'Six months?' I faced him side on. 'Is that all?'

'Pigs grow quickly. You don't want them getting much bigger. With boys, the taste is rich and succulent around now. But if you leave it until they hit puberty the meat can get a bit whiffy. At least that's what women say. It's called boar taint. Something to do with the hormones, but apparently men aren't so troubled by it. Can you imagine that? Sitting down for Sunday roast with your other half, only she's gagging as you carve and you can't see what the fuss is about?'

I felt as if I was learning something new with every observation Tom made. With my trainers ruined, and the hems of my jeans caked in mud I turned my attention to the wooden, tent-like structure in the corner of the paddock. Since that was what I had come to see.

'Now that is a proper pig ark,' I said admiringly.

Together, we set off towards it. 'Pigs sleep a lot,' said Tom.

'Throw in some fresh straw every few days and you can leave them to it.'

'It's that kind of thing I'm after. Only much smaller.'

Reaching the ark, Tom leaned one hand on the ridge and then rubbed the stubble on his chin. 'So, come on, what kind of size do you have in mind?

'I'm really not sure,' I confessed. 'I've only ever seen pictures on the internet. I've never seen a minipig in the flesh.'

Tom drummed his fingers on the wooden panel; thinking things through.

'I have to admit I'm intrigued,' he said. 'A couple of weeks from now I'm going to need an extra pair of hands to help round up my boys before their final journey. If you can help me out, I'll build you a minipig ark.'

This sounded like an offer I couldn't refuse. It meant I would now be presenting Emma with a quality dwelling for Butch and Roxi, tailor-made by someone who actually knew what he was doing.

'We have a deal,' I said, and struggled in vain to match Tom's vice-like grip as we shook hands.

6

Pride and Preparation

On the day my wife and kids were due to collect the minipigs, I had everything under control. As far as the family were concerned all they had to do was travel across several counties to pick them up. Everything else they could leave to me.

'It's my surprise,' I told Emma, as she placed the cat basket in the boot of the car. 'You'll love it.'

'It's certainly going to be an adventure,' she said, and kissed me goodbye. 'And your support means a great deal.'

As the car pulled away, I estimated that I had until late afternoon to get everything ready. The first thing I did was call Tom.

'We're on our own,' I said. 'It's OK to come up with the ark.'

'Understood. I'll load it onto the trailer right now.'

I had neglected to tell Emma about the repayment terms of my arrangement. I didn't think news that I would be helping Tom take his herd to slaughter would go down that well.

With the ark on the way, all I had left to do was prepare the space at the foot of the garden. It was already fenced for the chickens. Tragically, such measures hadn't been enough to keep Mister Fox out. Now, with Butch and Roxi on their way I figured we'd found the perfect deterrent.

'The day has come,' I said to our solitary hen on letting

myself in through the gate. Ever since the attack, Maggie had not been herself. She had lost her spirit as well as her tail feathers and barely acknowledged my presence. Her henhouse was purpose-built from plastic, looking much like an old-school Apple Macintosh, and came with a wire coop attached. Prior to our visit from the fox I had left the door to the coop open so the hens could come and go. Now, I kept Maggie contained, which I didn't like to see. 'You'll be free-ranging by sundown,' I assured her. 'Mark my words.'

What had been a generous enclosure for chickens finished at the back with a six-foot-high wooden fence. It divided us from our only neighbour. Roddie was a retired, slightly prickly widower who had lived in the village longer than anyone else I knew. Our houses were built in the same local brick and tile style, but the gardens could not have been more different. Having kids meant no matter how much care I put into our side, it would always be littered with 'outdoor' toys, which was basically shorthand for anything large, garish and plastic. Then there were our rabbit runs, and the devastating effect of the dog's nightly wee before bed. She tended to relieve herself in front of the shed. The result of such regular nitrogen hits meant the grass looked more like I'd been testing a flame-thrower on it. Roddie had no such issues. He kept his lawn looking like snooker baize, and screened us from his view with the cultivation of a small apple orchard.

We didn't see much of our neighbour, and not just because of the fruit trees. When Roddie ventured out, it would often be to attend parish meetings in a bid to make sure that the village wasn't going to be the site for a supermarket or six-lane superhighway. Despite his view that our community was under siege from developers, as well as itinerant tinkers and professional burglars, I always did my best to be a good neighbour. He could be abrupt at times, but Emma and I agreed this was down to living alone for twenty years and a fear of the unknown. Basically, Roddie was an old boy who liked things just as they were. For this reason, I had elected not to tell him about the new arrivals. Informing our neighbour that I had become a registered farmer just wouldn't go down well. I wanted the minipigs to settle in before we introduced them. That way, he would see for himself that they were no trouble at all.

In getting the enclosure fit for purpose, my main concern was the cat's grave. It had settled over time but somehow Misty still felt a little exposed. I had some spare flagstones behind the shed, and one of these proved the perfect solution. By the time I had stamped it flat, the sound of an old Land Rover pulling into the drive told me my masterplan for the minipigs was nearing completion.

'Well, what do you think?'

I found Tom standing proudly beside the trailer. The minipig ark was strapped to a pallet and mounted on the back. For a beat, I was lost for words. 'Wow!' I managed eventually. 'It's identical to yours in every way.'

'Except for the size, of course. I looked up "minipigs" online, and worked from a couple of images.'

As I stared at the fruits of his labours, Tom freed up the straps and then unloaded the ark. This he did by lifting it out by the carrying handle he'd attached to the ridge. I had expected something the size of a dog kennel. What he had built for me wasn't much bigger than a cardboard wine carrier.

'It's smaller than I imagined,' I said tactfully. 'Let's hope they get along.'

Tom swung the ark from the trailer and handed it to me.

'If they can fit inside a teacup, like all those pictures showed, they'll have plenty of room to stretch out inside here.'

Regardless of the scale, Tom had done a great job. The minipig ark fitted beautifully beside a spray of dogwood that grew just in front of Misty's grave. We even angled it in such a way that Butch and Roxi would be visible to us from the house.

Only Maggie the chicken seemed unhappy with the arrangement.

'What's got into her?' asked Tom. 'That's quite a squawk.'

For a hen that had been practically mute since the fox attack, Maggie was now making an almighty noise. She sounded very stressed, I thought. Enough for me to crouch before her coop and seek to calm her.

'Hey, what's up? You're safe now.'

As I watched over her, Tom set about clearing the space behind the shed.

'I can tell you now that this will be their toilet area,' he said eventually. Pigs are scrupulously clean when it comes to personal hygiene. They'll always relieve themselves as far from their shelter as possible.'

'That's good to know,' I said, as Maggie finally began to settle. 'I'm all for animals that look after themselves.'

'They'll even kick out a little straw from their ark so they can wipe their trotters before they step inside,' he added. 'Pig standards are impeccable.'

As I climbed to my feet, I noticed Tom's attention stayed with the hen.

'Old Maggie doesn't look too good,' he observed. 'A chicken shouldn't struggle for breath like that.'

Sure enough, with her eyes shut, Maggie appeared to be gasping.

'What's wrong with her?' I asked.

'Chickens are never the same after a fox attack. They're sensitive to any kind of disturbance. I'd give her some space. Let her get a grip while we check the perimeter.'

'For what?' I asked, following him to the back fence.

'Exit points,' he said, peering over the top of the fence, before crouching low. 'Your neighbour has an orchard back there. That's a big draw for any pig. I suggest we run some barbed wire from the foot of each fencepost to the next. It'll stop the little buggers from rooting underneath.'

'Won't they hurt themselves?' I asked.

Tom gave me a look. 'At the end of the day it doesn't matter how little they are. Pigs are pigs. You have a responsibility to keep them out of mischief.'

'I just can't see something that small doing anything drastic here.'

'Put it this way,' he replied. 'How would your wife react if a minipig went walkabout on your watch?'

I thought about this for a moment, before fetching a roll of barbed wire and a nail gun from the back of Tom's truck.

It was a two-man job. While Tom fixed the wire in place, I played to my strengths once again. By the time I came back out from the house with two mugs of steaming tea, he had completed the work. Without help, I'd have broken sweat, lacerated my hands and very possibly driven a nail into my knee. Tom had simply moved onto the next task.

'So,' I asked, on handing him a cup. 'What are we looking for?'

Tom was crouched low, tapping at a slat in the fence like a doctor might check a patient for tuberculosis.

'Any weakness that a minipig could use to its advantage,'

he said, before moving along to a gap at the foot of the fence. 'Like here.'

We were looking at the kind of hole you'd see in a Tom and Jerry skirting board.

'Surely not?'

'Oh, a pig can wriggle. When I open the gate, mine regularly try to storm their way out. You can't take any risks. I have some leftover wood in the trailer. Let's get this place minipig-proof!'

Tom was very thorough. He identified several potential exit points he believed Butch and Roxi could use. Privately, I didn't think any of them were big enough for a squirrel to slip through. Still, I had to place my trust in a man who had been rearing pigs for years. As Tom hammered lengths of wood in place, I handed him nail after nail and wondered how else I could pull my weight. At one point I decided to rake the grass, just to make it tidy. In the past, much to my annoyance, the chickens tended to scratch it away in places. Relatively speaking, compared to the rest of my garden, the chicken's enclosure was the one area that escaped most wear and tear. With Maggie contained, it had even begun to look quite lush and green. I glanced across at her pen, just to see how she was doing. It was then I realised that the grass was in better shape.

'Tom,' I said, and cleared my throat. 'Am I right in thinking Maggie has just passed away?'

For a man who had kept hens in his back garden for two decades, often dispatching the sick and the lame, Tom was very understanding about my loss. I wasn't distraught, just gutted that she'd missed out on the new protection measures and also a little bit mystified.

'You mustn't blame yourself,' he said, on crawling into the

coop to collect the carcass. Despite the size of his hands, Tom held her very gently. 'If our presence proved too much for her, think what would've happened had she woken up to find herself sharing the place with two minipigs.'

I considered this for a moment, specifically how the kids would've reacted on discovering our new arrivals in the company of a dead hen, and then nodded. 'Maybe it's for the best.'

'Take Maggie into the woods,' suggested Tom. 'You can consider it a sacrifice to the fox gods. It might keep them away from your new arrivals.'

'The minipigs were supposed to be a deterrent,' I said. 'It's a bit late now, though. Besides, I can't get rid of the body in the woods. After last time, the family would never forgive me. She was one of us.'

Tom looked a little bemused. 'So what do you propose? A memorial concert?'

'I just need to dispose of her respectfully.'

'How about burying her with your old cat?' Tom gestured at the slab in the corner of the enclosure.

I pulled a face. 'Not in Misty's Corner,' I said. 'The kids call it that as if she owns the spot. Besides, you can't get more than two inches deep without hitting thick clay.'

Tom sighed to himself. 'OK, how about *I* dispose of her respectfully instead? You can tell them Tom took good care of her.'

'Would you do that?' I asked, brightening now. 'They trust you.'

'Consider it done.' Tom turned to leave, tucking the chicken corpse under his arm at the same time. 'And good luck with the next round of animals.'

I watched him make his way across the garden.

'Tom,' I called out as he reached the gate. He turned to face me once more. I gestured at the dead chicken. 'You won't eat her, will you?'

Tom regarded me as if the thought had never entered his head. Then, with a smile, he left me alone in the enclosure.

That afternoon I kept myself busy by making the space look as good as I could. I cleaned out the henhouse and closed both doors. Then I found a painted metal birdcage in the shed and hung it from one of the overhanging lower branches of the oak. I even tied a welcoming red ribbon to the top of the new ark. I knew that I would have to tell the kids about Maggie at some point. This time, I hoped that the new pets would help them deal with the loss.

Once I'd laid out two bowls for food and water, all I had to do was head for the house and wait. With the enclosure up to speed I even decided I had earned the right to kick back and play videogames. It was a rare moment for me but I hoped the first of many. Once Butch and Roxi were commanding my family's time, I could reclaim the front room as my own.

Towards sundown, on taking a call from Emma to say she was five minutes away, I actually felt a hint of nerves about coming face to face with minipigs for real. I carried out some last-minute adjustments to the positioning of the ark, made sure the bed of straw I had laid inside was just so, and then hurried to the drive to meet them.

'Prepare to be amazed!' said Emma as she climbed from the car. 'They are the sweetest things.'

I opened the rear door to let the children pile out, and then peered inside.

'Where are you hiding them?' I asked. 'In the glove compartment?'

'Don't joke.' Emma kissed me on the cheek before making her way to the rear of the car. 'They'd probably fit.'

Eagerly, the kids and I gathered behind her. With great reverence, Emma opened up the boot lid. Inside, the cat basket now contained a lot of straw, which rustled when she lifted it out.

'How was the journey?' I asked.

'Fine. We didn't hear a squeak.'

'They are alive, aren't they? You know what they say about dogs and hot cars. I imagine the same advice applies here.'

'Don't fret. The girls have been checking on them regularly.'

Lou peered adoringly through the mesh. 'Oh, Dad! They're gorgeous. Their little snouts are like corks. I need to call Lauren, Jade, Darcy, Kath and Sara and get them over right away. Can they all stay for a sleepover?'

'Absolutely not. The minipigs need time to settle in. We need to offer them a calm environment.'

Emma lowered the basket so the little ones could see inside. 'Everyone is just excited right now,' she said. '*I'm* excited!'

I noticed May looking tense as Frank and Honey jostled for a view.

'I'm going to find Miso,' she said, and headed for the house. 'I'm worried he'll think I won't love him as much anymore.'

Smiling to myself, I realised I was feeling really happy about how things had turned out. Eager to impress them with the efforts I had made to make Butch and Roxi feel at home, I invited everyone to follow me into the back garden.

'Even if I say so myself,' I told them, heading for the enclosure, 'I've created a minipig paradise. They're going to love living out here. They really are.'

When I received no reply, I stopped and turned around. Nobody had moved an inch. They were just looking at me with puzzled, almost scandalised faces.

'What do you mean outside?' Emma clutched the cat basket protectively. 'Butch and Roxi will be living inside with *us*.'

7

Enter the Minipigs

In order to make a success of working from home, you need discipline and routine. You can't just shuffle out of bed when you please. You need to be washed, dressed and at your desk by a respectable hour, much like any other professional. Without some sense of self-discipline, you'd be in front of daytime telly before the week was out with biscuit crumbs all down your front.

Of course, there are chores to be done beforehand. While Emma power dresses upstairs, my task is to marshal the children into some sense of school readiness. Towards eight each morning, the kitchen is a scene of chaos. Once my family have cleared out for the day, I have only Sesi to deal with. So long as we don't encounter another dog, which runs the risk of bloodshed, I find a walk in the woods clears my head.

On the first morning following the arrival of the minipigs, that walk proved to be the calm before the storm.

Butch and Roxi had been with us for less than twenty-four hours. In that time they had been active for thirty minutes at most. After Emma had carried the cat basket in from the car and placed it on the kitchen tiles, the kids and I duly gathered around. Only May had kept her distance. She remained at the foot of the stairs, where Miso had stationed himself to see what the fuss was all about. Personally, I was still reeling from the ruling that the minipigs would be lodging under the same roof as us. As I awaited my first glimpse, I reasoned that they were so small we probably wouldn't notice them. It would be like keeping sea monkeys, I thought to myself.

'It's been a long day for them,' Emma had said, gently opening up the basket. 'Everyone stay quiet and I'll bring them out.'

Even with the mesh door open, all I could see was a bundle of straw.

'They are visible to the naked eye, aren't they?' I asked, to double check.

Shooting me a look, Emma positioned herself to reach into the basket. At once, something inside scrabbled and thumped against its confines. Whereas I would've snapped my hand clear, fearful of finding myself with just a bloody stump, Emma held her ground. She didn't lunge or make any sudden movement. She just kept her hand quite still as if to assure them she meant no harm. Evidently we had two very shy and nervous creatures in our presence. A tense minute passed. Finally, as a teenager with a rock-bottom boredom threshold, Lou broke the silence.

'Just tip them out,' she suggested.

'*Carefully*,' I stressed, in a bid to avoid a standoff between mother and daughter.

Still focused on the cat basket, Emma withdrew her hand and prepared to raise the back end. 'Shush now,' she said, lifting it an inch from the floor. 'Here we go.'

If I'd been surprised by the moment when the alien makes a break from John Hurt's chest cavity, this one prompted me to throw my hands in front of me. With no warning at all but for an ear-piercing squeal, two little blurs, one black and one pink with dark splodges, made a break for freedom.

'Watch out!' I cried, as they dived between the little ones. Both Frank and Honey yelped, and jumped apart faster than I had ever seen them move before. Lou was on her feet in no time, and then up on the chair behind her, shrieking in terror. Only Emma appealed for calm, but her voice was lost in the din.

'Where are they?' I spun around, ducking to look under the table, which is when I saw the pair properly for the first time.

In the corner of the kitchen, where the skirting boards met, two minute pigs were cowering nervously.

I knew they would be small. Nothing could've prepared me for the fact that the pair in front of my eyes were no bigger than kittens. Both had quivering snouts and eyes that locked onto me as I approached. I showed them my palms, as if that would help matters, and instructed everyone behind me to get a grip. 'Let's just stay calm,' I said, crouching at the same time. 'We have a situation, so let's not do anything foolish, OK?'

'What do you think they're going to do?' asked May from the steps. 'Shoot their way out?'

I turned to face her, frowning at the same time. Miso was at her side; clearly stunned by the sudden appearance of the two intruders. I watched his hackles rise, before he hissed and broke for the cat flap. At once, May was on her feet and glowering at me.

'See what you've done?' she yelled, before storming up to her bedroom. 'If Miso runs away I will *never* forgive you!'

I turned my attention to Emma. She was standing now, with the little ones clutching at her for protection.

'That went well,' she said. 'Could you have handled it any better?'

'*Me*?' I looked at Lou. 'What have I done?'

'Give them some space, Dad. You're freaking them out.'

I glanced at the minipiglets and then rose to my feet.

'Very well,' I said, showing my wife a '*was this wise?*' face, before retreating from the room. 'You guys do your bonding. I'll go talk May down from the ceiling.'

When it comes to any settling in period, it isn't the pets that need much attention. If anything, they just need to be left alone to get used to their new environment. In our household, it's the children that prove to be high maintenance. Sure enough, once Lou and the little ones had recovered their wits, they were keen to get as close as they could to Butch and Roxi. By the time I had assured May that the cat would slope home just as soon as he got hungry, I returned to find Emma's attention switched from the minipigs to the kids.

'You can look,' she instructed Honey and Frank. 'But you can't touch.'

The minipigs hadn't moved much. They were no longer huddled in the corner. Still, they were wary. Once again I could not quite compute what I was looking at. They looked more like toys than living animals. Part of me wondered whether a closer inspection would reveal battery compartments in their bellies. Emma was kneeling before them, restraining the little ones from venturing closer. Lou was standing over her mother taking pictures on her mobile phone.

'These are going straight to Facebook,' she said. 'I haven't even started with the video clips for YouTube.'

'The press call should wait,' I said. 'You can negotiate access rights in good time. Right now, Butch and Roxi's welfare is our priority.'

'Go and help your father collect the pig ark from outside,' said Emma, looking strained all of a sudden. 'Let's get them settled for the night.'

'Good idea. We should put them somewhere nice and quiet,' I suggested, without really thinking about where.

My office is located at the back of the house. Nobody else really comes in here. Generally, my wife and children just open the door, suggest I open a window for ventilation, and ask for money or lifts. As a result, finding myself in the company of minipiglets was a novelty in more ways than one.

'That's us clocked in,' I said to Sesi, having returned from our walk in the woods. Quietly, I eased into my chair. 'Let's earn a living.'

I didn't need to glance over my shoulder to be sure that now was a good time to start. I could see Tom's pig ark reflected in my monitor. Not only had Emma insisted that Butch and Roxi should live inside the house, she maintained that they were too young to be left unattended while she was away at work. Stationing them in my office was all her idea. I had drawn breath to protest, of course. My wife just spoke up before I could say a word, and in such a way that I could only sigh.

'They're as precious to me as the children. I couldn't do my job if I thought they were alone.'

With my writing document loaded on my computer, I squared up to my keyboard and prepared to begin. It was always my preference to work in silence, which is exactly what I had. Even so, with my fingers poised, I couldn't help but turn to check on the new arrivals. We'd placed their food and water beside the pig ark. Neither had been touched. Emma had also bought a litter tray. Following Tom's advice about their toilet habits, we had placed it in the far corner of the office. That, too, remained undisturbed, as did the blanket she had laid down on the advice of the breeder so they could have something to root around in. The only evidence of the minipigs' existence could be seen in the way the nest of straw inside the ark moved gently up and down. Butch and Roxi were sound asleep.

Exhausted by the upheaval, they hadn't stirred throughout the night. If these were Emma's surrogate offspring, I thought to myself, we were in for an easy ride. I glanced at the dog. Sesi was stationed behind the child gate to the boot room. With her ears pricked forward and her head tipped to one side, her attention was trained on the opening to the ark. She caught my eye and whined.

'*Shh*!' I hissed at her. 'You'll wake them.'

I couldn't blame her for wanting to come in. For a large dog to be presented with two tiny swine was quite a test of the senses. They didn't smell, as I had thought, but then only one of us had a canine nose. I also knew that Sesi liked a crispy pig's ear, which was a treat that would have to stop. Still, I imagined her taste buds were tormenting her just knowing what was hidden away inside the little ark. For this reason alone, the family hound would only ever be able to view the minipigs from behind bars.

With no appetite of my own to eat Butch and Roxi, I was free to get a little closer. I'd only been working for a couple of minutes, but my curiosity got the better of me. Having heard no sound from the ark, I was also a bit worried that Sesi's close proximity might've frightened them to death.

'Did nobody tell you about the dog on the journey here?' I was on my hands and knees, straining for a glimpse through the entrance. 'Hello? If you're alive, just blink or something.'

I heard nothing in response but the rhythmic rise and fall of the straw assured me they were still sleeping. I turned to Sesi once more. 'I wish you would spend more time sparked out,' I said. 'Maybe you can learn something from them about being low maintenance?'

Reassured that the minipigs hadn't perished in my care, I went back to my chair and set to work. I can't say I was entirely focused on the job. Sesi's continued whining didn't help matters. I knew that shutting the office door on her would only make things worse. To demonstrate this, you only had to look at the claw-marks that striped the panelling on the other side. She'd done the same thing to the back door, too. It was just her way of asking to be let in, even if it did leave the woodwork looking more like a grizzly bear had tried to gain entry. Eventually, with a chapter to write before I could take a break, I just tuned out as best I could

I had barely got through a paragraph before Sesi escalated the situation with a desperate sounding bark.

'I'm not listening,' I called back to her, while hammering the keyboard. 'You're wasting your dog-breath.'

It isn't unusual for me to remove my shoes before I sit at my desk. As a little ritual, I've undertaken it for years. Without footwear, I just feel more relaxed as I write. Pushing on with the words, however, I realised that something didn't feel quite right down there. I wiggled my toes, and took a moment to register that my socks were unusually warm. And very wet.

I drew back my chair to look down, and found myself facing the clenched rear flanks of a little pink pig. There she was, just in front of my feet, urinating prodigiously on the floor beneath my desk. Judging by the pool that my socks were now absorbing, it seemed like I had caught her just as she was finishing off. A little stunned, all I could do was watch and wait . . . and then wait some more. Had I started counting, I reckon I would've reached fifteen or maybe twenty before the stream of wee finally ceased. Then, with a shake of her tail, Roxi turned around and looked up at me.

'Hi,' I said brightly, and jabbed a thumb over my shoulder. 'For your information, the litter tray is over there. This is the wrong furthest corner from your pig ark. It's *my* corner, OK? The man's corner.'

Butch was close beside her, I realised. In fact, as she trotted around to inspect the back of my feet, he followed nervously behind. I wanted to move, given that my socks were entirely wet through. I just didn't want to startle them. Slowly, I reached down and touched her back. Roxi squeaked and flinched, but didn't disappear on me. Instead, when I used my fingertips to scratch her flank she seemed to like it.

'Wow,' I said, struck by how dense she felt for such a little thing. 'I won't ask what you're made from.'

Even as a child, petting pigs didn't feature on my list of things I was busting to experience. Still, as I stroked and patted Roxi, I felt quite touched that I had earned her trust so quickly. I had also surprised myself for not ranting at her for ruining my socks. Lifting one dripping foot after the other, I removed the offending items, rolled one inside the other and dropped it to the floor.

Immediately, Butch came alive. He broke for the sock ball, collected it in his mouth and trotted back to the pig ark.

'Hey!' I called out, as the dog went bonkers once more. 'Bring that here!'

The noise, combined with me spinning around in my chair, was enough to prompt Roxi into a retreat. Within seconds, she and her thieving brother were barricaded behind straw. I looked at Sesi, who appeared equally taken aback.

'What do we have here?' I asked. 'Minipigs or magpies?'

Having collected a fresh pair of socks and mopped up the

rest of the pig wee, I settled down at my desk once more. I didn't want to waste more time by trying to retrieve the old pair from the ark. As far as I was concerned, Butch and Roxi could have them as a comfort blanket. If they stayed quiet while I worked, that was fine by me.

Then the phone started to ring.

Before I'd even had a chance to reach for the handset, both minipigs rushed out squealing as if the call was coming from the slaughterhouse.

'This could be important!' I pleaded in vain. 'Pipe down!'

In the past, if the office phone rang when I was elsewhere in the house, Sesi would howl in order to alert me. She would also shut the hell up when I took the call. I didn't know enough about swine to say whether Butch and Roxi were trying to help me out. They just left me with no choice but to reach for the handset despite the fact that I could barely hear my own voice. Fortunately, it was my wife on the other end of the line.

'Is that the minipigs?' she asked. 'What on earth are you doing to them?'

'Nothing,' I protested, covering my other ear. 'The phone set them off, I think.' Emma didn't reply for a moment. She just listened.

'What's the name of that film you like?' she said next. 'The one when Burt Reynolds and his buddies go on that creepy canoe trip?'

'With the backwoodsmen?' I said, raising my voice to be heard over the squeals. 'That's *Deliverance*. And I know what you're going to suggest here.'

'It just reminds me of that scene.'

'Look, I'm not interfering with them sexually or anything like that. I'm just trying to get some work done.'

'God forbid you get a proper call,' she said. 'You'll have to do something about it.'

'Like put them outside?'

On the other end of the line, despite the noise, I swore I could hear the sound of Emma's nostrils flaring. 'Just don't distress them any more,' she said, before cutting me off in favour of a conference call.

As soon as I set the receiver back on the handset, Butch and Roxi fell quiet. I swivelled in my seat to address them. With two pairs of little eyes looking up at me, I decided that swearing at them would be inappropriate.

'So what was that all about?' I asked. 'Are you picking up on a frequency I can't hear? Is that it?' Neither minipig was in a position to answer me, of course. They just peeled away to bury their snouts into the folds of the blanket. 'You can pretend it's no big deal,' I continued, 'but I think we need some house rules. Number one, when the office phone rings, let's not make out that I've relocated to the barnyard, OK? Now, you're lucky that was Emma. At some point today, I'm expecting a call from a children's illustrator I'm hoping to work with. She's highly-regarded. Pretty famous, actually. This could be good for my career. Don't make her think I'm some kind of weird pig-interferer, alright?'

With nothing else to say on the subject, I left both minipigs under the watchful eye of my dog, and returned to work.

After lunch, things began to feel as if they might just settle down. Butch and Roxi had taken another long nap, and then ventured out once more. This time, they found their food and water. Like any parent, I have always made every effort to encourage my children to eat properly. The sound of someone chewing away with their mouth wide open can be hard to ignore at the best of times. Hearing the minipigs wolf down their pellet feed was different. The way they crunched away so enthusiastically was a distraction, but also heartening somehow. As for the long slurps that accompanied their drinking, I couldn't have made a better noise had I been presented with a bowl of soup and a straw. Having eaten, they both retired again for a while, before surfacing to take another toilet break. Underneath my desk once more.

This time, having removed a pair of wet socks for the second time, I kept them out of Butch's reach. It wasn't hard. I just placed them on the nearest surface that happened to be six inches off the ground. I also opted to put the litter tray in front of my feet in a bid to deter another flood. Once Emma got back from work, I decided, she could concentrate her efforts on toilet training them. My responsibilities extended no further than my work and the childcare. Even though my bare feet smelled of wee, my wife could deal with the cause. The minipigs were her lookout.

Throughout the afternoon, as Butch and Roxi explored the office, so my dog continued to bark and whine. I refused to let it bother me. I just told myself that Sesi would get used to the minipigs over time, and calm down accordingly. Butch and Roxi made very little noise. I could hear them shuffling around, spreading straw each time they left their ark, but the blanket seemed to keep them busy. Sitting cross-legged in my chair, just as a precaution, I was free to crack on without worrying about their welfare.

Maybe I became a little too absorbed in my work. All I can say is that I didn't register the silence until a couple of minutes after it fell. When I realised the dog had ceased making any noise whatsoever, my fingers froze on the keyboard. It just wasn't a good sign. In the woods, if Sesi stopped making any noise, it usually meant she had found a badger's carcass or an old deer bone. With a gasp, I switched my focus from the document on the screen to the reflection of the space behind me. I saw no sign of the minipigs outside their ark, and spun around with a start.

'I really don't think that's wise,' I said, scrambling to my feet. 'Guys, you should come away from there.'

While my back was turned, the minipigs had found the courage to direct their attention towards the large white wolf behind the gate. Butch had held back a little bit, unlike his sister, who was now in front of the child gate, directly underneath Sesi's gaze. Up on her feet, every muscle in the dog's body was braced to pounce. Thanks to the barrier between them, I was surprised but not alarmed in any way. Most of all I was curious to see what would happen. Certainly Roxi showed no hint of fear. She simply scratched her flank with one hind leg, before looking up into a pair of hungry, lupine eyes.

Then, with a friendly squeak, and much to my horror, the little pink and blotchy minipig calmly slipped between the bars. Such was her size she barely brushed the edges.

'Stop!' I cried, and tried in vain to grab her by the tail.

From the dog's point of view, a light pork snack had just offered itself up on a plate. I could only think Sesi was too stunned to act, for Roxi calmly trotted through the space between her forelegs. A moment later, having lost the minipig from her line of sight, my White Canadian Shepherd performed a one hundred and eighty degree switch. Roxi seemed entirely unconcerned, despite finding herself right between Sesi's front paws once more. The dog looked down at her, and then across at me as if seeking my permission.

'Don't harm her,' I pleaded in a whisper. One wrong move to rescue Roxi could prompt the dog to pounce. This was undoubtedly the first time Sesi had encountered a living creature that wasn't remotely scared of her, and that included humans. I looked at Roxi, inspecting the tiles with her snout, she seemed entirely unconcerned.

Then the first string of spittle fell upon the minipig's back, followed by another, and that's when my huge hellhound gave in to temptation. A cry died in my throat as Sesi's head dropped down, but she didn't snap up Roxi in her jaws. Instead, very gently, she licked her. Not just once but slathering her all over. For a moment, I wondered whether maybe a dog's saliva had the capacity to tenderise its prey. I just could not fathom any other reason why she would be doing it. I pictured Emma returning home at the end of a hard day's work, only for me to explain that one of her minipigs had been softened up and then prematurely butchered in the worst way possible. At this point, Butch decided that he was missing out on the action. Having passed clean through the bars, he stood obediently beside his sister and allowed himself to be washed down as she had. On seeing this, I approached the child gate as calmly as I could.

Sensing my approach, Sesi stopped licking them. She glanced around. Then she growled at me.

Freezing on the spot, all I could do was watch helplessly as two glistening little creatures lifted their heads in glee at the attention. I had always considered my dog to possess a murderous instinct, not a maternal one, but here it was in full effect. For the minipigs, who had only recently been taken from their mother, Sesi certainly seemed to meet a need. Within minutes, all three animals had nestled on the dog's bed. That's when I reminded myself to blink and breathe.

For a short while I just looked on, lulled in some way by such an unusual but touching arrangement. Finally, the office phone rang out. I wasn't alone in being startled by the sound. I just didn't feel the need to join in with the cacophony of piggy squeals and canine howling as I rushed back into my office to pick it up.

8

In-House Training

After just one working day it was clear to me that mixing business with Butch and Roxi might not work out as Emma had hoped. If they wanted to pass the probation period, and continue to share office space with me, my minipig colleagues would have to get their act together.

Turning the space under my desk into a toilet wasn't a great start to our professional relationship. Nor was kicking off throughout my telephone calls. When the illustrator rang I really should've explained the source of the noise straight away. I just didn't know where to start, and so I said nothing about it. Of course, given that their squealing made it sound like I was hanging out with two psychopaths suffering simultaneous asthma attacks, she seemed hesitant and baffled. All I could do was hope she'd conclude that I was one of those people who worked to experimental music, and agree to illustrate one of my books.

It was the last time I ever heard from her.

On the upside, Sesi had proved to be a positive distraction. The fact that she didn't appear to want to maul or eat little pigs was both a surprise and a big relief. Having invested so much cash, sweat and tears in the dog's training, I now hoped our new arrivals would look up to her and even learn a thing or two about obedience, respect and devotion to their master.

What I didn't consider, until Emma returned home, was the fact that Butch and Roxi had already identified who was in charge around here.

'Who let the minipigs out?' my wife demanded to know, on walking into the front room to find them playing with the little ones. 'Have you not been watching them?'

'It's the child gate bars,' I explained. 'As soon as the kids came home from school, they both just slipped through to see them.'

'What about the dog?' she asked, with some alarm.

'Relax,' I said. 'Unless she learns how to use a jemmy, there's no way Sesi can beat those bars.'

With a sigh, Emma shrugged the bag from her shoulder. 'You know what I mean,' she said, searching inside it next. 'Sesi is a sweetheart with the family, but no dog can be trusted one hundred per cent.'

'I think the same rule applies to minipigs,' I said, as Butch and Roxi continued to delight our youngest children. Roxi jumped and skittered around her brother, who stood there flicking his tail as if preparing to charge. 'I'm not suggesting they could turn on the kids like one of those pitbulls. I just mean when your back's turned they can be a little bit naughty.'

'They just need training,' said Emma, and promptly produced a pack of nuts from her bag. 'Have you taught them any tricks yet?'

'Oh, sure,' I replied, trying not to think about today's wasted man-hours. 'Butch is a regular David Blaine. You should see him levitate.'

Ignoring this, Emma opened the pack and tapped a couple of nuts into her hand. I had assumed they were for her, until she crouched and offered them up to the minipigs. I watched them scamper eagerly towards her. 'You see,' she said. 'They just need a little encouragement.'

'That's bribery,' I replied, as she held the treats away from their snouts. 'Besides, what tricks are you hoping to teach them?'

'Sit!' shrilled Emma. 'Minipiggies, *sit*!'

I glanced at the little ones. Like me, they were finding this request a bit hard to take in. Frank even sat down himself, as if that might satisfy his mother's request.

'Emma,' I said hesitantly. 'You're going to need a lot of nuts.'

'I'm not expecting them to sit straight away,' she said, feeding them the treats from the palm of her hand. 'You just have to get them used to commands.'

'OK,' I said. 'Perhaps we can start with something more useful. Like encouraging them to use the litter tray, rather than the floor under my desk.'

'Let them get used to their new home, first,' replied Emma, folding the pack of nuts back into her bag. 'They're smart creatures. They'll soon work out how to fit in with the family.'

Clutching the bag in one hand, my wife then rose to her feet and left us for the kitchen. Led by their snouts, Butch and Roxi followed close behind.

As any parent knows, children under the age of six don't put themselves to bed. Persuading them to recognise when their waking hours are numbered is a daily mission. One that begins a long, long time before lights out. For us, the process involves slowly winding down Frank and Honey for the day. Post supper, that means nothing that might hype them up, such as blue sweeties or television programmes featuring extreme skateboarding or Richard Hammond. Ideally, by the time we'd brushed their teeth and read them a story, all the little ones would want to do is curl up and close their eyes.

To be fair, at their age I'd have behaved in the same way

at bedtime had my parents invited two minipigs into my life.

'Let go of the door frame, Frank,' I pleaded, struggling to escort him back to bed for the third time. 'You don't need to keep saying goodnight to them. The same goes for your sister.'

'They'll still be here in the morning,' Emma assured him. She had just been upstairs settling Honey when Frank popped down again. Losing patience this time, I had turned him around at the threshold of the front room, which is when he demonstrated an early understanding of non-violent resistance.

'I'm not so sure,' I said, prising my son, one finger at a time, from the door frame. 'This is just added grief, Emma. All I want to do is sit down in front of the fireplace with a nice glass of wine.'

'You see,' she said, addressing Frank now in a quiet but firm voice. 'Daddy drinks because you're being difficult.'

Oblivious to the unrest they were causing, Butch and Roxi continued to potter about the front room. Just then, my concern was in getting Frank back to bed. Freeing him from the door frame, I narrowed my gaze at my wife.

'Under the circumstances,' I muttered, as I slung my young son over my shoulder to carry him upstairs. 'Daddy deserves a break.'

Ten minutes later, following a repeat of the bedtime story, I returned to the front room to find Emma waiting for me with two glasses of wine on the coffee table. She smiled just as I did.

'Success?' she asked brightly, from her place on the sofa.

'Sure,' I said, and kissed her on the forehead, 'but one glass is enough for me. You have the other one, eh?'

Lou had taken the place of the little ones by now. She was fussing over Butch and Roxi, both of whom were thoroughly enjoying being so spoiled. All three were gathered on the hearth, basking in the heat from the fire. I did worry about what might happen should the minipigs cook themselves. I just assumed I would pick up on the succulent smell of danger before it was too late. On the television, with the volume muted, the credits were beginning to scroll for our regular soap opera. As long as the older girls had done their homework, we enjoyed unwinding together. It was one of those family moments in which nobody spoke a word to one another. Usually, I would break it with a comment dismissing the acting quality or a shaky script, only to be roundly shushed.

'Give your sister a shout,' Emma asked Lou. 'It's about to start.'

'*May*!' she bellowed, ignoring the countless number of times I'd asked her not to interpret the request so literally, before taking herself to sit beside her mother on the sofa. 'She'll be down in a minute.'

Maybe the minipigs had already become used to such close attention. They certainly didn't like being abandoned one bit. Just as soon as Lou had settled in her place, Butch and Roxi rushed to her feet.

'Cute-o-rama!' sighed Lou, and reached down to pat their heads. 'Do you guys want to sit here, too?'

'Don't you think we should have some boundaries?' I asked, but my reservations fell on deaf ears.

With squeaks of what could only be happiness, both minipigs allowed themselves to be lifted onto laps. Lou and Emma took one each, but quickly Butch and Roxi were padding over the cushions; seeking out any crumbs they could find, it seemed. I had yet to take my place on the sofa, despite the fact that the programme was starting. Just then, however, it didn't look like an entirely restful place to settle. Certainly Emma looked unsettled.

'We can't watch it without sound,' she muttered, and joined the minipigs in searching down the back of the sofa. 'Has anyone seen the remote?'

I looked around, but saw no sign of it. Then I considered who might have had it last.

'I think I know where to look,' I said, and headed for the office.

Sure enough, along with a balled up pair of socks, still wet with stale urine, I found the missing control unit inside Butch and Roxi's ark. From the straw bedding, I also hauled out a toy car and a drying up cloth. Lastly, I collected Sesi's prize chew. The dog was at the child gate, watching me retrieve Butch's stash.

'Here you go,' I said, tossing the dog chew back over the gate for her. 'That's one mad minipig we've taken on here. Please don't punish him for taking it. I'm sure once he realises the risk he ran it'll never happen again.'

Leaving Sesi to gnaw on her chew, I crossed the kitchen with the remote control in hand. I noticed that the rubber edging and some of the buttons had been nibbled. I was also aware that a fair bit of straw had been dragged across the kitchen tiles. It was for this reason, when May stopped me in the hallway, that she found me muttering to myself.

'Have you seen him?' she demanded. 'Where the hell is he?'

Aware that I had just been caught swearing under my breath, I chose not to ask her to be quite so confrontational. My cat-loving daughter was standing in front of the door, arms folded, glaring at me.

'Have I seen who, sweetheart?'

'Miso. He always sits with me while I do my homework.'

I didn't like to point out that he only ever hung out with her until his food had gone down. An hour or so after his supper, he'd disappear on us in favour of another family. Instead, I glanced at the stairs.

'Has he touched his water?' I asked, counting four brightly coloured cups stationed randomly on different steps.

'Dad. He's missing. We need to go out and search for him!'

'Let's just relax,' I said, and motioned at the door. 'Come in and watch the television. I think Mum and Lou are hoping to turn the minipigs into soap opera fans.'

'The minipigs are the problem!' she snapped back, only to look at me pleadingly. 'Miso is *scared* of them.'

'May,' I said, and placed a reassuring hand on her shoulder. 'Miso isn't entirely happy about anything. His hackles rise when the post comes through the door. The other day your mum found him yowling on the landing. It turned out there was a spider on the stairs. He wouldn't go down until she'd removed it.'

'This is different,' she stressed, and I realised she was close to tears.

Still clutching the nibbled remote in one hand, I offered my daughter a hug. 'He'll get used to the minipigs,' I assured her. 'We all will,' I added, but mostly to myself. 'Come on, we've missed the start already. If the cat hasn't shown up by the time it's over, I promise I'll look for him.'

In the front room, May tucked herself in on the sofa beside her mother. I knew that Emma must've overheard the conversation as she drew her close for a cuddle. Lou was still playing with Butch and Roxi. Both minipigs had settled on the seat beside her; the spot traditionally reserved for me.

'So you found the remote?' observed Emma. 'Let's have some sound, then.'

I pointed the device at the telly, and pressed the button in question. Even before the sound failed to rise, it didn't feel right to touch. If anything, the button felt completely missing.

'Great,' I muttered, inspecting the remote. 'Looks like Butch had a particular taste for the volume, contrast and subtitle options.'

'Never mind that now,' said Emma. 'Just turn it up at the telly and sit down.'

I'd only ever used the remote to operate the television. It meant I had to put up with several seconds of sighs and tutting until I found the buttons hidden behind a panel on the side of the box itself. Butch didn't look terribly sorry for the trouble he had caused. Even Roxi seemed unsympathetic. She certainly wasn't particularly willing to create some room for me on the sofa.

'Dad!' complained Lou, as I eased myself in at the end. 'You're squashing her!'

'She'll move,' I said, despite what sounded like grunts of stubbornness. Eventually, aware that I might well be about to cause some harm, I encouraged the minipig to shift by placing her in the space behind her brother. In response, as if disgruntled that she couldn't see the screen from back there, Roxi threw her head back and snorted angrily. I couldn't help but laugh. That something so tiny could become so cross was most amusing. She then scrambled from the spot I had designated for her, crossed my lap and attempted to rout me. 'What *are* you doing?' I asked, watching her shovel her snout between my thigh and the side of the sofa. 'Give it up, missy. I'm staying put.'

'Pipe down!' hissed Emma, without turning her attention from the screen.

If Roxi's temper took me by surprise, her determination to push me out of the way was something else. Had she possessed the strength to match it, I might well have lost the fight. Instead, with great gentleness, I employed the minipig as an elbow rest and just let her get on with it. The experience wasn't a comfortable one. Midway through the programme, I had to concede a little space simply because in digging down she'd corkscrewed around to such a degree that her rear trotters were planted in my groin. By the time the credits rolled, I was ready to admit defeat.

'Miso is still missing,' said May, as I rose to my feet. I watched Roxi unplug herself from the side of the sofa and then settle in the space I had warmed for her. 'Can we look for him now?'

At this point, the only thing I had in mind was what we might cook for our supper. I looked at Emma. Like Lou, she was gazing adoringly at Butch and Roxi.

'I still can't believe we have pigs in the living room,' she said.

'Me neither,' I said, a little darker than intended. 'So, what are we going to do about Miso?'

'Look outside,' May suggested. 'On the lane.'

'There's no need to think the worst,' I assured her.

Emma looked across at me. Some concern crept into her expression. 'It might be worth checking.'

Lou's anxiety levels were beginning to rise. I could see the tension in her face.

'When did you last see him?' I asked.

'When we all got home from school,' she replied. 'He headed outside straight away.'

In a normal house, you'd expect to find the cat flap installed in the back door. What stopped us from doing this was the dog. Even the bravest of felines would think twice about popping inside if they were going to step right into a gated room occupied by a massive hound with fangs. It meant we had to rethink the entry point. The only practical alternative

I could come up with involved providing access through the French windows in the front room. Breaking out a glass pane and replacing it with a Perspex flap in a plywood frame was not a task I carried out with relish. I didn't carry it out at all, in fact. Tom did it for me. The finished work was professional, a pity, and a permanent reminder that some pets never play well together.

That evening, with Lou looking more agitated by the minute, I decided the first thing I should do was simply call for the cat. I didn't really want to step outside unless I absolutely had to. It was cold, damp and dark, after all. So, I got down on all fours in front of the French windows, peeled away the curtain and pushed open the flap.

What I saw caused me to catch my breath. Fortunately, I kept it together for the sake of the girls. I just hadn't expected to find Miso staring right back at me, but there he was; outside the house, sitting on the other side with his wraithlike features illuminated by the light through the flap. That it looked like he'd been stationed there for some time didn't rattle me as much as the look in his eyes. I have never seen such unbridled resentment before, and it told me just one thing. As far as the cat was concerned, the minipigs had pushed this household's pet population to the very limit of his patience. I couldn't be sure what was going through his mind, but I had a feeling he would spend the night stalking wild rabbits in vengeful fury. Closing the flap, I rose to my feet and assembled a smile.

'He's fine,' I said. 'Just getting some air.'

9

How Much Trouble Can They Be?

Having a baby for the first time is a life-changing event. It's as exciting as it is exhausting, while everything they do is a source of wonder and fascination. The novelty wears off just a little bit with the second child. Their arrival is still a cause for celebration, of course, and you love with them with all your heart. Even so, you're an experienced parent now. With the exception of my wife, you can't go into denial about the sacrifices ahead. For me, with every newborn that followed, so the workload we faced stopped me from welling up at the sight of a pair of scratch mittens.

I'm sure if we were childless then Butch and Roxi would've caused me to coo at least once. Instead, in the weeks after the minipigs bundled into our lives, I basically just ran around after them. Being so young, of course, they took regular naps, and that's when I squeezed in my work. Emma was always happy to pull her weight each evening, but by then the pair were so tired they were less of a handful. They also took instruction from my wife, as well as all the nuts she used to reward them for doing little more than blinking or staying upright. It meant that when the day arrived for me to repay Tom for building their ark, frankly I was relieved to be out of the house.

Emma and the kids were out at the time, but I was hardly

in a position to enjoy the solitude. As it was a Saturday, they had headed for the nearby town so Lou could buy some designer bag that she'd been saving for. At fourteen, my eldest daughter was looking forward to a birthday sleepover at a friend's house that evening. Apparently, Lou's last bag was now so unfashionable it couldn't even be donated to the charity shop in case anyone realised it had come from her.

With the minipigs left in my care, again, I had decided to wait for Butch and Roxi to turn in for their first snooze of the day before stepping out. Finally, once they had settled inside their ark, I left Sesi to watch over them and made my way down the lane. This time, I wore a pair of boots and a pair of trousers that I didn't mind getting muddy. I knew that Tom was around. His Land Rover was backed into the track that led to the stables and his fields. He'd also attached his stock trailer to the tow bar. It looked like a stunted horsebox, with slats in the side for ventilation. Tom had already fitted the ramp. I peered inside. With straw on the floor, only one thing was missing.

Feeling a little strange about the task ahead, I made my way towards the pig field. I heard Tom before I spotted him. He has a big voice to match his frame. At first I thought he must have company. I stopped at the paddock fence and looked beyond the horses. I could see him chatting away under the eaves of the stranded tree. The trunk obscured my view of whoever he was with. When the pig finally shifted into my line of sight, I realised this was a one-way conversation. Tom's tone stopped me from joining him straight away. I couldn't hear what he was saying, but it sounded gentle and soothing, and quite clearly some kind of farewell. As he spoke, he fed the great beast pieces of apple, which he deftly cut up with a penknife. When the other two pigs emerged from the treeline, Tom produced another apple from his pocket. Eventually, as he coaxed them over, I figured I should make my presence known. On reaching the gate, I coughed discreetly.

'Hey there!' Smartly, Tom stepped away from the pigs. 'Good to see you reporting for duty.' The sheepish look on his face reminded me of one of my little ones whenever they denied raiding the biscuit jar despite the chocolate around their mouths. I let myself in through the gate, and did my best to pretend I hadn't heard a word. Shaking hands, Tom sized me up and down. 'How are things?' he asked, his smile fading. 'You look tired.'

'I'm fine, really,' I stopped there, unable to break from Tom's gaze.

A look of concern crossed his face. 'What's up? You can tell me.'

I swallowed uncomfortably, feeling like someone who'd just dialled an emotional support helpline but couldn't quite find his voice. 'It's the minipigs,' I said in a small voice. 'I had no idea what it would involve!'

Tom sighed. I couldn't tell if he was looking at me in sympathy or with pity. 'We've got time for a brew,' he said, and invited me to follow him.

Pulling up a straw bale, I sat with my mug of tea cradled in both hands, and declined the offer of a digestive. We were inside Tom's workshop. He had built it against the back of the stables, with a window on one side overlooking the paddock. Here, in this inner sanctuary, he performed the kind of DIY Dark Arts that I could only dream about. On the wall in front of me hung all manner of tools, from rivet guns to clamps and spanners, punches, files and rasps. The workbench, littered with wood shavings, screws and sketches of gates and trellises, was at least six inches thick. It was worn at the edges, stained in places most likely by sweat from Tom's brow, and scarred with countless nicks and grooves. My desk at home just didn't compare.

'You know what?' I said, as Tom leaned back against his bench. 'It's been a couple of weeks since Maggie died, and still nobody has noticed. I feel like Butch and Roxi have taken over our lives.'

Tom considered this for a moment. 'Pigs should be low maintenance,' he said. 'How much trouble can they be?'

'I don't know where to begin. Every time I sit down to work, there's an issue.'

'Not with the ark, I hope.' Tom took a slug from his mug. 'We've had some rain recently. Sometimes that causes fresh joints to warp a little. I can always pop up and straighten things out if it's a problem.'

At this, I found myself struggling to look him in the eyes.

'There's no danger of the rain getting into the joints,' I said quietly, and looked at my wellies. 'The ark is inside. Butch and Roxi are living in my office.'

I didn't need to see Tom choke on his tea. When the droplets hit the floor in front of my boots, preceded by a volley of spluttering, I knew just what his response had been. 'You have the pigs in your house? *Pigs*?'

'Minipigs,' I said, as if that would make it more acceptable. 'It wasn't my decision. I had assumed they'd be in the garden. Everything was ready for them out there.'

Tom turned his attention to the window. Over in the field, his three pigs rooted and grazed oblivious to the fate that would soon befall them.

'I think perhaps this should be our secret,' he said eventually.

'If people in the village find out, you'll never hear the last of it. You can't keep livestock indoors. It isn't natural.'

'Butch and Roxi seem to love it,' I said. 'So do Emma and the kids, only they don't have to deal with them during the day *and* hold down a job.'

'Are they digging up the carpets, then?'

'That's about the only thing they haven't done,' I said. 'But it's only a matter of time. My biggest problem at the moment is the toilet training.'

'Really?' Tom sounded surprised. 'Aren't they going as far from their ark as they can?'

'They are now,' I said, 'but for days I was cleaning up under my desk. I put down a litter tray but both minipigs ignored it. Then they got adventurous, and started going elsewhere. It took me a while to work out where. The smell gave it away after a while.'

'So, where are they taking themselves?'

'The far end of what they must consider to be their enclosure,' I said dolefully. 'In the front room. Behind the telly.' For a moment, we sat facing one another without a word. I could've shown him my hands, dried and cracked from constantly wringing out a cloth in a bucket of water mixed with disinfectant. I just didn't want Tom to think I was making a fuss. 'Emma keeps saying they'll soon work out what the litter tray is for. Short of urinating in it myself, I'm at a loss as to how to make them realise it's not there for them to root around in. Then there's their enthusiasm for food,' I added, feeling a sudden desire just to unload all the issues I'd had to handle.

'Pigs certainly like to eat,' said Tom.

'They're supposed to have two meals a day. They get a cup of pig feed when I start work, and another when I finish. It's just I can't open the fridge to fix myself something to eat without finding them at my feet. Even if I creep into the kitchen when they're asleep, their hearing is so sharp that I can expect a minor stampede within seconds. What's worse is that I've been reading all the paperwork that came from DEFRA. It's illegal to feed them kitchen scraps, but Butch and Roxi don't know that. They just never give up. Tom, I don't want to go to prison over some baby sweetcorn.'

'You need to be strong-willed. Ignore them.'

I placed my cup on the worktop surface. 'You can't ignore a minipig,' I told him, thinking about what had become a nightly tussle with Roxi for my place on the sofa. 'If I could, they wouldn't be a problem.'

'How are they with the dog?'

'Great. Same with the children. I can't say the same for the cat, though. They've seriously freaked him out. Miso will never be the same again.'

'I'm sure he'll get used to having them around.'

'No, he won't,' I insisted. 'Emma took him to the vet for his annual boosters the other day. He practically turned himself inside out with stress just as soon as she shut him in the cat basket.'

'Would that be the same one she used to transport Butch and Roxi?' Tom smiled quietly to himself. I knew what he was getting at here.

'Yes, I know that cats are sensitive to smell, and I realise I probably should've scrubbed it clean before we used it again. At the time, I just didn't think it would be a big deal for him. I'd brushed out all the straw, and laid down a clean sheet of newspaper, so it wasn't like we didn't care. Anyway, Miso must've made an impression on the vet because Emma came home with a prescription for anti-panic medication.'

'For the *cat*?'

'Once a day with food.' I paused there, aware that what I had said might take a moment to process. 'Tom, Miso's on tranquillisers now. Quite a high dose, apparently.'

Tom shook his head. 'Man, he must've been anxious.'

'Not any more. We rarely used to see him. Now, much to Lou's relief, and also what's left of the wild rabbit population, he just doesn't go out. It's disturbing, though. His eyes are like saucers and he stares at the wall all the time. If it wasn't for Butch and Roxi, Miso would still be devoting his time to looking down at us. It isn't the same anymore.'

Finishing his tea, Tom rose to his feet. 'Once my boys are loaded up, I'll run you home. It's about time I saw these minipigs for myself.'

Half an hour later, Tom lifted the ramp on his stock box and secured it shut with the steel pin and chain. I stood beside him, still holding the two wooden boards he had given me to help herd his pigs. It hadn't been as testing as I thought. Tom had pretty much led them in using chunks of apple as encouragement. What really struck me was that the closer they got to the open box, the quieter he had become.

'Everything alright?' I asked, as he toyed with the vehicle keys in his hand.

'Sure,' he said, clearly surfacing from his thoughts. 'This is never nice, but it's all part of keeping pigs.'

'Don't let Emma hear you say that,' I said, and headed around for the passenger door.

Settling in the driver's seat, Tom slotted the key in the ignition, but didn't turn on the engine straight away.

'Tell me something,' he said, turning to face me. 'What is it that draws your wife to these animals? As I see it, there's no better taste than pork you've reared yourself, but for Emma that's unthinkable. So, what's the big attraction?'

I thought about this for a moment. 'I know they've been a hassle,' I said, 'but I can't deny they have character. Roxi is remarkably bold, while her brother brings up the rear and steals anything that isn't nailed down. Emma thinks it's endearing.' I looked at my lap for a moment, reflecting on what this meant. 'I suppose you could say she just really values the good things about them.'

Tom nodded to himself, as if he could identify with her outlook.

'Those boards,' he said next. 'Maybe you should borrow them. If you're going to learn to love your minipigs, it sounds like they could use a little direction.'

'There's only one direction I'd like them to go,' I said as he gunned the engine into life. 'Outside.'

We pulled out into the lane. I could hear the pigs jostling in the box behind us. I was glad not to be travelling all the way with them. Then again, on reaching my house, a little bit of me wished we could've just driven straight on.

'What's all that about?' Tom killed the engine.

From inside the house, the muffled yet intense and altogether alarming sound of squealing could be heard. I faced Tom. He looked appalled.

'Seems there's a phone call for me,' I said. 'It's probably best if I let the answer machine take it.'

The telephone had stopped ringing by the time I opened the back door. Inside, I found Sesi lying on her blanket with her head on her paws. Instead of launching into her usual overenthusiastic welcome, she just turned her eyes up at us.

'She looks a little cheesed off,' observed Tom. 'Surely she hasn't met her match in these minipigs?' As he spoke, Roxi and then Butch squeezed through the bars from my office to greet him. They circled his legs enthusiastically, evidently hoping he had food. Tom looked at me, his mouth open a little bit. 'They're *tiny* pigs.'

'Actually, they've grown a little bit since we got them,' I said. 'Roxi more than Butch. It won't be long before she gets stuck between the bars.'

Clearly taken aback, Tom crouched down for a closer inspection. 'Look at you two,' he said. 'Now I see why Emma's fallen for them in such a big way.'

'You haven't spent time with them,' I replied. 'Mark my words, they'll test your patience.'

Tom seemed not to hear me. His focus was entirely locked on the minipigs.

'Hello, handsome,' he cooed, scratching behind Butch's ear. 'Is that good, little man? Is it?'

'Tom!' I said, battling for his attention now. 'Whose side are you on here?'

With a chuckle, he rose up to his full height. 'OK, I can imagine they're a handful.' He peered into my office. Butch and Roxi had hauled yet more straw across the floor around my chair. Just inside the ark lay a child's shoe and a DVD case. 'Have you taught them to do that?' he asked. 'I've heard of pig-keepers who encourage their herd to nose a ball around. Hoarding stuff is new to me, though.'

'Butch even stole my cash card the other day,' I grumbled. 'He'll be making withdrawals next.'

Rasping at his stubble with one hand, Tom looked down at the two creatures settling with the dog on her blanket. Roxi had just flopped over into Sesi's plush white coat, while Butch was trying to tuck himself between her forelegs. The dog continued to peer up at me like all this was my fault.

'Let's hope your minipigs don't outgrow the ark,' said Tom. 'Else poor Sesi is going to find herself looking for a new bed to sleep on.'

'Roxi is bound to get a little bit bigger,' I said, 'and so will Butch. But that's cool. We're ready.'

I wasn't sure I sounded as confident as I hoped. Tom just continued to stare at them. 'It's remarkable how different they look to each other.'

'I can tell you why,' I said, a little brighter this time because at last this was a pig topic I did know about. 'Butch and Roxi are the result of combining different breeds over several generations. It's been carefully done so only the smallest qualities shine through. Such different markings just show their lineage.'

'I see.' Tom looked at me. 'So, they're kind of mongrel pigs.'

'*Mongrels*?' I caught my breath, feeling a little insulted all of a sudden. Butch and Roxi didn't belong to me, but I'd been forced to shell out a small fortune for them. Even so, now that Tom had said it out loud, I realised it was a fair point. 'OK, so these minipigs aren't a pedigree, but that might change for future generations. Maybe in time they'll even breed out their thieving tendencies.'

'And fondness for soiling your front room.' With his hands in his pockets, Tom observed the two new arrivals. 'You have yourself a pair of poppets in these pigs, but you wouldn't catch them living under the same roof as me. Not with the noise I just heard them making. I'd have them outside, where they belong.'

'Yeah, but your wife is a reasonable woman,' I said. 'Emma has made her decision. As far as she's concerned, Butch and Roxi are as much a part of the family as, well . . . me.'

Tom didn't reply straight away. He found the keys to his

Land Rover and let me reflect on what I'd just shared with him. 'You need to man up,' he said eventually. 'You can't work in this environment, let alone live in it. You're unhappy with the situation. That's obvious. If you let it pass these pigs won't let you forget it. They'll be a constant reminder that you crumbled.'

Tom turned for the door and stepped out onto the yard. I watched him glance at the sky, presumably predicting in that moment the exact forecast for the coming days. 'Once I've let the ground recover,' he said next, 'I'll be looking for some more pigs for my field. Why don't you come to auction with me? I think it's important you get some insight into this world you've joined. It'll be packed with traditional breeds, and you're encouraged to check them out at close quarters. You need to know if you're bidding for a pig that's healthy, after all, or a boar guaranteed to deliver the goods if breeding is your game.'

'Sure,' I said, trying not to sound baffled by the invitation. 'Sounds interesting.'

'Oh, it is.' Tom held out his great hands as if preparing for a catch. Then he levelled his gaze at me. 'It's all in the balls,' he said and squeezed his fingers together. 'You have to get down and give them a really good feel. If they're firm, and not too spongy, that's the pig you want to pick. It really is the surest way to find out if he's got what it takes.'

Leaving me at the threshold, Tom told me to stay lucky before heading back to his vehicle. Grim-faced, I closed the door before turning to face my dog, Butch and Roxi.

'We might have a problem,' I said with a sigh.

10

More Than a Match

When questioned about the balance of power in our relationship, my wife has a very clear view. To anyone looking in, it may well appear that Emma is firmly in charge. Between us, she maintains, it is always me who has the final say on matters.

In my experience, that's usually, 'if you're sure' or 'I'm going to walk the dog'.

That afternoon, as I cleared up after the minipigs, I resolved to make some changes. Even fixing a sandwich turned out to be a trial. Our fridge door always opened with a slight squeak of the hinges. To Butch and Roxi's bat-like ears, it was the equivalent of sounding the lunch bell.

'I'm not giving you anything from the kitchen,' I told them. 'Prison scares me.'

Crossing to the table with my plate in hand and a glass of water in the other, the minipigs seized their chance. It was the kind of ploy you would expect from a small, annoying yappy dog. Two minipigs, arguably smarter than a canine and operating as one, left me with no chance. First Roxi timed her move so that she would be directly under my foot as I stepped forward. Perhaps knowing that I would take emergency measures to avoid squashing his sister, Butch then positioned himself a little further ahead. It left me nowhere to go. All I could do was stumble. I stayed upright, but it was enough to cause my sandwich to slip from the plate.

'Don't touch it!' I yelled, already fearing what I would tell the lags in the showers when they asked me what I was in for. '*Stay back from the sandwich!*'

Roxi fell upon it with her brother close behind. Undeterred, I wrenched it from their jaws and raised it over my head. The minipigs responded by going nuts at me. Roxi repeatedly head-butted my ankles while Butch just bellowed and stamped his trotters. I even found myself backing off from them a bit. I realised this might've been foolish, as Tom had said that pigs could smell fear. I wasn't sure that an ankle-high version could cause much damage. I just thought it was best not to provoke them, at least until I understood them better. Even so, I was taken aback at how furious a minipig could be simply because something didn't quite go their way.

'OK, enough! Calm down!' In a show of defiance, I tipped my sandwich into the bin. Next I showed them the empty plate. 'Here's the deal. Next time I go to the supermarket, I'll buy some fruit and we'll keep it all in a treat tin on top of the woodpile. So long as it doesn't come into the kitchen, I'll remain a free man. Until then,' I stressed, 'I expect better behaviour than this.'

Butch and Roxi had calmed down a little since I got rid of the sandwich. As they were now sniffing around the base of the bin, I doubted very much they were aware of my bid to regain some authority. 'Am I making myself clear here?' I demanded to know. 'Don't make me send you to minipig camp, guys. I'll set one up myself if they don't exist. All I ask is that you behave like normal pets.'

It was then, on hearing a growl from behind the child gate, that I realised we were not alone in the kitchen.

I couldn't say for sure whether Miso had been sitting under the table throughout this time. Maybe Sesi had only just picked up on his presence. Even if the cat had floated in midway through my reprimand of the minipigs, all the evidence suggested he was totally oblivious to what was going on around him.

'Hey, puss. Everything alright?' I was aware that Miso *never* came into the kitchen. Doing so only placed him in the dog's line of sight, which is exactly what had just happened. In response, and perhaps most striking of all, not a single hair on the cat's back was standing on end.

Miso looked just as you might imagine, which was basically a cat whacked out on prescription tranquillisers. For one thing, he was facing the table leg. Getting down on all fours, I drew level as gently as I could. Being this close to him was weird for me. Normally, he'd run just as soon as I called his name. This time, Miso didn't blink or twitch his whiskers, even when I waved a hand in front of his face or whispered 'raaabbbiiitt' in his ear. For once, I felt some concern for him. I was about to scoop him up just to check him over when the phone began to ring.

By now I was used to having to raise my voice over the squealing when I picked up the handset. This time was no exception.

'Hello?' I said, and then covered the mouthpiece to plead with Butch and Roxi.

'Mr Whyman?' asked the voice on the other end of the line.

'*Shhh*! Speaking.'

'Roddie here.'

As soon as my neighbour identified himself, I stopped slouching and cleared my throat.

'What a nice surprise. How are you?'

Crooking the phone between my shoulder and ear, I attempted to shepherd two noisy, marauding minipigs out of the room.

'Mr Whyman—'

'Please, call me Matt.'

'Mr Whyman, I've been meaning to call you . . .' He trailed off there, as I feared he might. 'I know this sounds a little unconventional, but are you keeping pigs in your house?'

'Pigs?'

I offered him a feeble chuckle.

'Well, are you?'

'No,' I replied, without thinking. 'Of course not.'

I had managed to close the door on Butch and Roxi. Even so, it hadn't done much to calm them down.

'I have no wish to pry,' said Roddie, who clearly wasn't going to settle for my abrupt denial. 'It's just I keep hearing the same noise.'

I felt both stupid and panic-stricken. There was no reason for me to cover for the minipigs. All the paperwork was in place. I had legitimate farming credentials. What persuaded me to lie to my neighbour was very simple. I was just too embarrassed to admit what we had got ourselves into here.

'What kind of noise?' I asked, grasping for any way out of this conversation. At the same time, I opened the door by a fraction and glared at both minipigs.

'The squeals are unmistakable, Mr Whyman.'

'It could be the boiler,' I suggested, grimacing to myself as I heard the words leave my lips.

'It's definitely pigs. There are pigs inside your house.'

I stared through the window, aware that my excuses were going nowhere.

'Sorry,' I said. 'Did you say pigs?'

'Yes.'

'Oh! I thought you said . . . *figs*.'

'Figs?'

I closed my eyes, wishing a large hole would open up, preferably under Butch and Roxi. They didn't like being shut out one bit. Clearly the door did little to block whatever frequency from the phone was upsetting them.

'You're right. We do have some little pigs here actually. They're pets.'

Roddie didn't respond straight away. I braced myself for him to complain about the noise, wondering what I could say in their defence. Finally, much to my surprise, my neighbour began to chuckle.

'That's absurd,' he said eventually. 'I've never heard of such a foolish thing in my life. A pig as a pet, indeed!'

Despite hating myself for attempting to dodge the truth, I actually felt a little bit indignant at being laughed at like this.

'They make *great* pets,' I insisted. 'So easy to house train.'

'But not so easy to keep quiet, Mr Whyman.'

At last, it was quite clear just why he had called.

'We're working on it,' I assured him. 'They only kick up a fuss like this when someone is using the telephone. At any other time you wouldn't know they're here.'

'Then perhaps I should ring off,' said Roddie. 'If it means that awful racket will stop.'

'It will,' I said. 'And I'm sorry if it's bothered you. Peace and quiet will be restored very soon. For your sake as much as mine.'

That afternoon, even with the family still out shopping, I felt the need for a little space. This wasn't easy to find. Every time I moved from one room to another, I would be followed by minipigs. Even retreating behind the Saturday paper didn't stop them. Personally, I would never have encouraged them to use the sofa. Roxi, in particular, was beginning to seriously challenge me for a comfortable spot I had always considered my own. The third time she attempted to rout me, I gave up on what had been a really interesting article about emigrating in middle age. Instead, I removed a page and folded it over Butch's back. For the next minute or so, I amused myself by watching him scamper around looking like a very small bull that had smashed his way through a tent.

It was the first occasion that day that I had smiled, I realised, which didn't actually make me feel much better. Still, it left me thinking perhaps my time would be more constructively spent connecting with our new arrivals, rather than rueing their presence in my personal space.

'Pay attention, minipigs! Let's see how you really measure up.' Moving the litter tray to one side, something we had left in front of the television in the hope that they would start using it, I switched on the Playstation and the screen. Then, collecting not one controller but two, I turned to face the minipigs. 'Who wants to be first to take me on, then? According to Emma, you guys are supposed to know what to do with these things. I guess it's just a question of working out where your strengths lie. So, what's it to be? We have shoot 'em ups and beat 'em ups, but unfortunately for you no dig 'em ups. We do have some sports games, though. How about soccer? Tom said pigs like a kickabout, so I'm guessing you might know the rules.'

Some minutes later, at seven nil up, I concluded that the research Emma had cited was flawed.

Not only had Butch and Roxi both failed to pick up the second controller, let alone sink one into the back of the net, they were showing little interest in the game itself. My performance didn't go unobserved, however. After driving home the ball from one hundred yards, straight over the head of the minipigs' stationary keeper, Miso elected to join us. This, I only realised when I felt his fur brush my elbow. I glanced around to find the cat sitting right beside me. I still found it hard to get used to such intimacy with him. He seemed oblivious to Butch and Roxi, who had dragged the throw from the sofa and were busy rooting under it. At first, I thought the cat's eyes were locked on the game. It was the dilated state of his gaze that told me it was more likely to be focused on an imaginary point midway between him and the screen. Miso seemed content, in a zonked out way, so I went back to playing the game. It was a bit pointless, of course, but weaving around players that seemed asleep on their feet was mildly distracting. If I had kept my eye on the ball, in the really important sense of the word, I might've stopped the second controller from going missing.

As soon as I noticed that it had disappeared, I knew just where to look. The rest of the family easily forgave Butch's pilfering. I just wished he would learn to stop doing it. I had reached the kitchen when I saw him. He was beating a hasty path out of my office. I crouched so I could somehow express my displeasure. This would be an opportunity to begin the training, I thought to myself. Unfortunately, our little black minipig was in too much of a hurry to oblige. He rushed around Sesi, slipped through the second set of bars and scampered between my legs. I faced the dog. She looked as puzzled as I felt.

'What's got into him?' I asked, and headed for the office.

It didn't feel right. As soon as I reached into the ark and laid my hand on what I had come to retrieve, I knew that

Butch had made his mark on the videogame controller. The thumb stick, designed to steer everything from soccer players to supercharged cars around the screen, was now just a plastic stump. Of the four buttons employed to attack, defend, accelerate, brake, or restart the game when things had gone really badly, only two were left. The controller was a write-off, and my mood had soured considerably.

'This has got to stop!' I snarled, on marching back to the front room. It didn't matter that this was the only time the second controller had ever been used. Nobody else in the family showed any interest in videogames. Like an increasing number of men of my generation, it was a form of escapism I had never left behind. In my mind, the second controller existed for the day that my son was old enough to join me in tuning out from life's stresses and strains. In a way it represented a rite of passage. Now a minipig had destroyed it, and I was hopping mad. '*Butch*!' I called ahead, 'you and I are about to fall out!'

What I found in the front room stopped me dead in my tracks. At once, all the anger coursing through my system melted away. In its place, I felt a surge of elation.

'Look what you've done!' I declared. 'You *clever* boy!'

As I walked in, Butch was just stepping off the litter tray. The deposit he had left behind was a joy to behold.

'This,' I said, beaming as I pointed at the dark and shiny solid, 'is a turning point. From here on out, we can work together to get along. You're a good minipig, Butch. Now, where's your sister?'

I looked around, keen to find her so she could learn from her brother. Miso hadn't moved from his spot in front of the television but Roxi was nowhere to be seen. The football match remained frozen where I had paused it. At least it did for the next second or so, before a fizzing sound marked the moment that the screen turned black and the console lights died. Then, as if aware that she was in big trouble, Roxi made a break from behind the television. This time, she hadn't soiled the newspaper I had taken to laying out for them. The only thing of note back there, when I went to investigate, was the cable she had just chewed clean through.

11

An Embarrassment to the Family

Returning from shopping with the kids, Emma's first response on seeing the murderous look on my face was to ask after the welfare of the minipigs.

'I've just seen Tom's wife,' she added before I could reply. 'Apparently you helped him round up his pigs for the slaughterhouse.'

'I was returning a favour,' I admitted. 'He stopped in here on the way, in fact.'

Warily, my wife looked around. 'So exactly where are Butch and Roxi?' she asked with some concern.

'You'd better sit down,' I said. 'It's bad news.'

'What?' Emma paled visibly, only for the colour to return in a different way when I explained the fate that had befallen my gaming console. 'We can get a new cable,' she said. 'The minipigs are irreplaceable.'

'The minipigs are fine.' I waved a hand dismissively in the direction of the office. 'They're asleep in their ark now. They're really tired after the day we've had. My guess is they'll be out for the count until the next person dials my office line.'

With the children back in the house, our attention was duly turned towards picking up coats, sorting out minor squabbles and fixing squash drinks. While May took herself off to talk to Miso, presumably in the hope that her voice might coax him back to his senses, the little ones headed into the front room to make a den. Whatever they constructed, whether it took them into outer space or the high seas, somehow it would always require a titanic tidying operation afterwards. Lou remained in the kitchen with us. She was looking very pleased with herself. I assumed this was down to the shopping bags she carried inside. When I asked to see what she had chosen, my eldest daughter seemed at a loss as to what I was talking about.

'Never mind about that,' she said eventually, and dug around in one of the bags. 'Wait until you see this!'

I was familiar with the magazine she produced. Her mother had been buying it on a monthly basis for years. It was packed with celebrity news and gossip, all of which left me cold. Unless nobody was around, of course, in which case I'd find myself having a quick flick through just to be sure that I still had grounds for disapproval.

'Whatever you have to show me,' I said, as Lou searched through the pages, 'I imagine I won't be impressed.'

'Prepare to eat your words, Dad.'

Having found what she was looking for, Lou laid out the magazine and invited me to take a look. I glanced at Emma. She suggested I do just as Lou asked. With a sigh, I glanced at the gallery of paparazzi pictures that accompanied the article in question. I wasn't going to admit that I recognised every single actress and pop diva featured. Then I realised each one was cradling a small pig, and snatched the magazine into my possession.

'"The latest must-have accessory,"' I said, reading snatches from the article out loud. '"This season, no A-lister can afford to be seen without a minipig."' I looked up at Emma and Lou. 'That's a lot of famous farmers,' I pointed out.

'Do you know what this means?' said Lou, and clapped her hands like a seal on the side of a show pool. 'At school, everyone knows me as the minipig girl. It's so cool. Now I'm going to be just like the celebrities!'

For a moment I looked long and hard at my daughter. 'A minipig is not just for a fashion parade,' I told her. 'As much as it pains me to say this, a minipig is for *life*.'

'Looks like we got Butch and Roxi at the right time,' said Emma. 'I imagine this means their asking price will rise.'

'Also the selling price,' I replied. 'Hey, we could make a quick profit here!'

'Dad!' Lou sounded outraged. 'Don't even joke about things like that.'

I didn't like to say I was being serious. Instead, seeing that we were talking about the minipigs, I figured now was a good time to raise the issue of their living arrangements.

'Actually, I've been thinking about Butch and Roxi's welfare.' I took a breath. 'I know they're small and cute, and cutting edge now, it seems, but they're still pigs. And a pig's natural environment does not involve lolling about on the sofa waiting for the lottery results. They need to be outside, like any other pig.'

I had predicted that my opening pitch would be met with some resistance. Lou began tutting and rolling her eyes as I spoke. Emma just folded her arms.

'They can't go outside,' she said. 'That would be cruel.'

'How is it cruel?' I turned out the palms of my hands, shrugging at the same time. 'They can still come in, when the time is right. Having said that, my guess is they'll want to stay outside because frankly that's where they belong. In fresh air.'

'We'd be reported,' added Lou. 'Kicking them out of the

house would be so heartless. What's wrong with you, Dad? You're *sick*!'

Having gently suggested Butch and Roxi might be more comfortable under the oak, my wife and eldest daughter now left me feeling like I should be banned from keeping animals for the rest of my natural life.

'Let's be realistic here,' I said, hoping to start again. 'Since they've arrived, my work has gone down the tubes. Several important contacts think I'm operating out of a barn and our next door neighbour has made a complaint about the noise.'

'So how is throwing them out going to help?' asked Lou.

'They won't hear the phone from outside,' I said. 'And I'll be free to earn a living again.'

With her arms crossed still, Emma considered my proposal. I was close to getting her on board. I could feel it in the air. As my wife drew breath to speak, Honey scuttled in from the living room.

'What's the matter?' asked Emma, for it was clear she had something to tell us.

Honey pointed at where she had come from. I realised what she had found.

'Is it the poo?' I asked. 'I'm aware of the situation.'

'What kind?' Emma looked at me quizzically.

'Well, it isn't mine!' I protested.

'It didn't even enter my head,' she countered. 'Until you said it.'

Together, with Lou, we turned to face Honey. Her brother was close behind, looking horribly like he had prodded their discovery with his finger. Honey beamed at us.

'It's minipig poo,' I said to clarify. 'A minipig poo in the litter tray.'

The grin that eased across my wife's face told me just what this meant.

'A housetrained minipig,' she said eventually, 'deserves to be under the same roof as us.'

Washing up after supper, I looked at my reflection in the taps and did not like what I saw. The fact that my face appeared to have been shaped by a spoon was down to the curve of the chrome. It left me looking chinless, which was pretty much how I felt.

'Show some balls,' I muttered at myself. 'What are you? A man or mouse?'

Peeling off my Marigolds, I turned and looked around. I'd cleared the plates and cups from the table. The next chore was the straw on the floor. What with all the coming and going, the minipigs had seriously begun to spread it around the house. Nobody else seemed bothered by it. I just kept sweeping it up thinking pretty soon we'd all be rooting around in blankets and grunting at each other. As I cleared up the mess, grumbling at the dog, Lou came downstairs looking all dressed up.

'The sleepover started half an hour ago. Everyone will be there. Can you give me a lift in five minutes, please?'

'Fashionably late as ever,' I said. 'Sure I can.'

Lou beamed at me. 'Great. I'll just finish packing my stuff.'

As she took herself back to her bedroom, negotiating the cups laid out on the steps, May herself descended. She was carrying Miso in her arms. The cat's eyes were wide open, but his focus was strikingly slack.

'Everything alright?' I asked, as May placed him gently before his food bowl.

'Miso needs his tea,' she said, crouching over him. 'Come on, pussycat. Eat something.'

The cat didn't move, even when May picked up a nugget of food and waved it in front of his face.

'He'll eat when he's ready,' I said, aware that May was looking very tense indeed. 'There's no need to force it down his throat.'

'If I don't,' she said, waving another nugget, 'he'll die.'

I frowned at her response. 'That's a bit extreme, isn't it?'

Finally, she turned to face me.

'Miso is on hunger strike,' she said. 'Because of *them*.'

I didn't have to ask if she was referring to the minipigs. That she couldn't even bring herself to mention them by name spoke volumes.

'He's just getting used to the medication,' I said. 'Things will calm down for him.'

'Calm down?' May returned her attention to the cat. 'He's traumatised! Not eating is his only way of protesting. I know him, Dad. I understand what makes him tick.'

As she spoke, I realised that I might have found an ally in my bid to relocate the minipigs.

'Would Miso be happier if they lived outside?' I asked.

May nodded solemnly. Upstairs, I could hear Emma had begun the uphill task of encouraging the little ones to bed. From experience, the only conversation that I could have with her at this time would centre upon whether we had a bottle of wine in the house. If I was going to win out on where to put the minipigs, I would need to make a watertight case. Aware that May was resting her hopes on me now, I promised her that I would do what I could.

Five minutes later, all ready to drop off my eldest daughter at her sleepover, I found myself sitting alone in the car. I had the engine running as well as the wipers. A light drizzle was falling. As soon as we stepped out of the house, Lou had turned back because she wanted to find an umbrella. It was no surprise this simple task took so much time. At her age, leaving the house always took several attempts before she had remembered everything. I had listened to an entire news bulletin before Lou reappeared. Dashing across the headlights, my eldest daughter arranged her bag on the back seat and then climbed in beside me.

'Where's your umbrella?' I asked.

Lou looked at me blankly.

'Oh!' she said eventually. 'Sorry. My mobile rang when I was looking for it. Sophie wants me to bring my new make-up. By the time I found that, I'd completely forgotten about the umbrella.' She glanced out of the window. 'Never mind about that now, though. It's eased off a bit.'

'You haven't even got a coat,' I pointed out. 'You can't go out dressed like that.'

'Dad! It's just a girly sleepover. Can we go now?'

I didn't need to ask for directions. I regularly ferried Lou to and from friends' houses in the village. Now she was getting older, the times for dropping off and picking up were eating into the evenings. A sleepover was always a result. It meant I didn't have to sit around, clock watching with a sparkling water in hand. The payoff was that eventually Lou's turn would come around and our place would be turned into a hell mouth populated by shrieking teens. Until that moment arrived, however, I could look forward to enjoying a night of peace and Pinot Grigio.

'You can drop me here,' said Lou, as we neared our destination. 'This is fine.'

'Eh? You want to walk? You never walk anywhere'

'Really,' she said, sounding a bit desperate. 'It's no problem.'

I pulled up on the side of the lane. The house was just beyond Tom's smallholding. It was only as she reached for the car door that I realised what was happening here.

'You're embarrassed to be seen with me,' I said. 'That's what all this is about, isn't it? You don't want me to take you to the door.'

Awkwardly, Lou turned and asked me to consider what I was wearing.

'Just look at yourself,' she said.

'A sweatshirt and drawstring trousers,' I said. 'What's wrong with that?'

'A stripy sweatshirt,' she pointed out. 'And *chequered* drawstring trousers.'

'But it's the weekend. They're comfortable.'

Lou bunched her lips. 'It isn't just the trousers or the pattern clash,' she said. 'You're in those plastic clog things.'

I glanced down at my footwear. I agreed they didn't look great. They were just easy to slip on, which made them ideal for a quick trip down the lane and back. Lou's central problem, I realised, was that I had one foot in a clog belonging to me, and the other in one worn by Emma. Mine was green. Hers was pink. I looked back at Lou. She flattened her lips in a way that reminded me of her mother.

'Dad, you can't go out dressed like that.' Climbing out of the car, my daughter collected her overnight bag from the back seat. 'Promise me you'll drive straight home.'

In the front room, I found Emma browsing at her laptop from the sofa. The television offered the usual Saturday night fare: some brash, colourful talent show, which probably explained why she'd muted the volume.

'Can I ask you something?' I stood in front of the screen,

facing my wife, and spread my arms. 'Do I look like an embarrassment?'

With her fingers dancing over the keyboard and the laptop light illuminating her face, Emma responded without actually looking up. 'In what context?'

I spread my arms. 'That's the wrong answer,' I said, and told her what had happened. 'I accept that mismatched footwear wasn't great, but it's not like I'm wearing a comedy tie or anything like that.'

At this, Emma glanced up from her laptop. She shrugged, and then returned to her work. 'I suppose not,' she said.

Realising that I would get more response from Miso, I huffed irritably, swiped the remote control from the top of the TV and parked myself on the sofa beside her.

'What are you browsing?' I asked, and began flicking through the channels.

'Just stuff about minipigs,' she said. 'The usual.'

I made no effort to look for myself. Instead, I settled on a documentary about truckers in lumberjack shirts journeying across some icebound wilderness. Just at that moment, it seemed like I might have found a chance to unwind. After the day I'd had, indeed throughout the weeks since Butch and Roxi arrived, this was a rare moment. The little ones were asleep, Lou was away for the night and May was upstairs talking to Miso, accompanied, no doubt, by the imaginary beep of a life support machine. With my wife at my side, and real men on the telly, this was my idea of family life. I might have had to get up to sort out the volume, admittedly, but that was a minor issue.

As I returned to my seat, the minipig responsible for destroying so much of the remote control function trotted in to join us. Butch sniffed around a little, and then flopped in front of the fireplace as if he'd just received a dart from a blowpipe. He didn't bother me, soil the carpet or attempt to steal anything. Nor were demands made for food or water. Placing my feet up on the coffee table, I settled back and prepared to start my Saturday night at home.

I can't say how many minutes passed before I gasped like a surfacing free diver and jumped to my feet. One, maybe two or three. All I know is that until that moment my mind was wandering a little. In the TV documentary, one of the truckers was driving along with his trusty hound sitting on the passenger seat beside him. He was talking about how his dog would lay down his life for him, and that he would do the same for the dog. I was just dwelling on whether I should feel guilty for not feeling the same way about Sesi when it occurred to me that one of our pets was missing.

'*Roxi*!' I said out loud, when Emma asked me what was wrong. 'Where is she?'

'Around,' she replied, seemingly unconcerned. 'She's probably asleep.'

'She can't be,' I said. 'That minipig is hell bent on claiming my space on the sofa. Every time I sit here, she tries to turf me off!'

Emma folded her laptop shut. I knew that I had made a valid point. I could see it in the way her eyes tightened at the corners. 'Surely not?' She rose from the sofa, and then her attention locked on a point behind me. 'The cat flap!'

Sidestepping Butch, who was spark out on the carpet, she rushed to the French windows

'It would be a struggle for her to get through now,' I said, mindful that Roxi was growing quite a bit faster than Butch.

Emma opened up the flap and peered outside. 'It's just she's such a determined little soul.'

I was about to agree, but something in her comment made me stop and think. When she turned to face me, I had only one thing to add.

'So too is your eldest daughter,' I said gravely, for I realised just what had happened here. 'And I know where we'll find your minipig.'

Understandably, the mother of Lou's friend was surprised to see me at her door. I don't think my weekend clothing ensemble was entirely responsible, nor the fact that I was about twelve hours too early. It was the look on my face that informed her that a small matter I needed to discuss with my daughter couldn't wait until the next day.

'Dad? What are you doing here? Oh, my God, *Dad*!'

I don't make a habit of walking in unannounced on teenage girls' sleepover parties. This was a first, in fact, and very much the last because what I found confirmed that Lou was about to be grounded for at least a decade. Before I could utter a word, however, I noted her friends sizing me up and down.

'You were right,' said one girl under her breath. 'That's a *mad* look.'

I was facing approximately a dozen of Lou's friends. I didn't count them. My attention was locked on the minipig they were pampering on the bed. Roxi was stretched out on her side, eyes closed and snuffling blissfully at all the attention she was receiving. Lou looked at me aghast.

'She must've fallen asleep in my bag,' she said weakly, and then appeared to give up on the worst excuse ever. 'I'm in big trouble, aren't I?'

Unwilling to create a scene in front of her friends. I told her that the three of us were leaving. 'I'm sorry to spoil the party,' I said, and hefted Roxi off the bed, 'but this girl is too little to be out late.'

With the minipig under one arm and Lou protesting at her exit behind me, I turned and marched from the room. The mother of Lou's friend must've been trying to listen in because I found her near the top of the stairs. The poor woman couldn't have been aware of her extra houseguest, however, because seeing Roxi under my arm caused her to gasp and cover her mouth. I just didn't know how to explain what was going on here. So, I simply smiled at her as we edged by and wished her luck with the rest of the sleepover.

Back at our house, Lou rushed straight for the arms of her mum. There on the sofa, dabbing at her eyes with a crumpled tissue from the car, she confessed what she had done. Before I'd had a chance to park my plastic clogs and join them, Roxi headed straight for my spot. Her brother quickly hopped on beside her. There, the minipigs arranged themselves nose to tail, as if settled for the night.

Even if I had wanted to claim my place, I was too angry to do anything other than stand to deliver my lecture.

'In this country,' I began, 'by law, if you want to take any kind of swine anywhere, no matter how big or small, then you have to apply for a movement licence. It lists all the different reasons for transporting your pig, and I believe you have to tick one. So, if you want to take it to market, export or slaughter, you do the paperwork first.' I paused there to compose myself. 'Lou, *there's no box to tick for taking a pig to a sleepover*!'

'OK, I think she knows she's done wrong.' Emma gave Lou a hug, and kissed her on the forehead. 'The main thing is, Roxi is safe and well.'

'I'm just embarrassed,' mumbled Lou, shredding her damp tissue.

'And rightly so,' I replied, feeling a little calmer now I had made my point.

Lou looked up at me. 'No, I mean about you stomping in without warning.'

'Dressed like that?' Mortified, Emma turned to Lou. 'Baby, that must've been awful. Give it a while. Your friends will move on, I'm sure.'

I drew breath to point out that at least I had taken the trouble to slip on matching green clogs this time but thought better of it. Had I known this whole sorry event would lead to my humiliation, I'd have made my entrance dressed as a fright clown. Not by ringing the front door bell. I'd have climbed up the drainpipe to knock at their bedroom window.

I felt cornered, close to defeat, and yet something deep inside me refused to just let it go. I still needed to take a stand here. I may have given in several times on the subject that was uppermost in my mind. Now, I would be laying down the law. At that moment, fired up by my failure to be taken seriously, I realised I did indeed still possess the one thing that Tom had made me think might be missing. Two, if we're being pedantic.

'We've crossed a line here,' I said, struggling to contain myself. 'Ever since these minipigs arrived, my feelings about their place in this family have been ignored.'

'Oh, that isn't true,' said Emma. 'I know you've worked incredibly hard to take care of them during the day. We really appreciate it. Don't we, Lou?' she added, digging our daughter in the ribs in order to prompt an agreement. 'They're supposed to be fun pets, not a source of stress.'

I touched my temple with my fingertips, feeling a headache

coming on. There was no backing down for me now, however. I knew Emma and Lou wouldn't like what I had to say, but I was going to tell them anyway. For I had balls now. Big ones. If I was a male pig at market, Tom would surely pick me.

'I'm just trying to do the right thing for us all,' I started. 'If that's going to happen, changes have to be made. If we let the minipigs run riot, just as they have been, next thing they'll learn to climb the stairs and then there'll be no escape.' I rolled my shoulders, commanding silence from the sofa. 'Tomorrow morning, and this is the last word on the subject, Butch and Roxi will be moving to the end of the garden.'

Butch getting under my feet.

Nothing like a tummy tickle.

Watering can attack.

PART TWO

Out for the count following the vet's visit.

Itchy itchy, scratchy scratchy.

Keeping a watchful eye.

12

A Drunk with a Digger

A great weight was off my shoulders. It had settled there on the day the minipigs moved in, and grown steadily with every deposit they left behind the television. We may have cracked that issue, with Butch and Roxi both using the litter tray, but that was not enough to make me comfortable. They needed to be outside. It was the only way to stop them from overtaking our lives. At last that was going to happen.

When I had laid down the law, Emma looked set to dismiss me. What stopped her was Lou. In a bid to overturn my decision, she piped up that Maggie the hen might be too fragile to handle their presence.

'She's used to being alone now,' Lou had offered. 'Bundling minipigs in with that poor chicken could be too much of a shock for her. Do you really want that on your conscience?'

'Lou,' I had said, and cleared my throat. 'We've been a chicken-free family for quite a while now.'

I knew there was no way they could come back from this. All I had to do was inform them that Maggie was dead already, an event that nobody had noticed, and both my wife and eldest daughter fell silent.

I slept well that night. With no more worries on my mind, I had curled up beside Emma and quickly crashed out. Unusually, we weren't awoken by the little ones within seconds of daybreak. Despite the blackout curtains in the room they shared, the sun only had to crack over the horizon to persuade both Honey and Frank that they should be bouncing on their parents' bed. When I did stir, facing the clock on my side, I was heartened to see that they had left us alone for a whole extra hour.

Emma was still asleep behind me. I could tell by the rhythm of her breathing. Every now and then, she would snore just ever so slightly. It was nice to hear, in view of the fact that I was always the one being jabbed in the ribs for disturbing her. I had no intention of complaining. I was still practically out for the count. Instead, I rolled over, rested my arm across her shoulder, and nodded off once more. It proved to be that weird phase in a sleep cycle, when your dreams are most active and unsettling. Mine involved finding myself trapped inside a minipig ark. In such cramped conditions, the air was hot, musky and suffocating. I didn't panic. I just lay there appealing for help. I must've called out for real at one point, because it served to wake me up.

'Ugh,' I croaked groggily, 'that was awful.'

Emma just grunted in response. I lay there, my hand on her shoulder still, thinking that we really must get up in case the little ones tried to drop lighted matches down the back of the television or some such. Then my wife grunted again, loud enough this time for me to open my eyes wide. I looked across the pillow, and then jumped out from under the sheets.

'Emma?' I whispered, scrambling to find my pants. '*Emma*! Where the hell are you?' In bed still, half dressed in a negligee belonging to my wife, Roxi the minipig eyed me contentedly. 'I didn't touch you,' I said, backing into the corner of the room where I struggled into my jeans. 'Nothing bad has happened here!'

As I said this, the bedroom door opened and I found myself facing my family.

'Matt?' Dressed for the day, Emma narrowed her eyes at me. 'What's going on?' She turned round and instructed the little ones to stay back. 'Don't look,' she told them. 'It'll never leave you.'

Lou and May peeped in beside her, and recoiled in horror. 'Oh, gross!'

Roxi didn't seem upset at all. She seemed quite comfortable, in fact. I looked at my wife, feeling panic-stricken, and that's when they all started laughing.

Had I not been caught still pulling up my jeans, I might've seen the funny side. As it was, I waited until I could hear myself again, and informed them all that this prank had gone too far.

'Whose idea was it to dress her up?' I asked, and quickly found a t-shirt. 'If it was anyone other than your mother, I'm going to be stopping allowances.'

By now, still giggling in fits, Emma's eyes were starting to shine. 'Oh, lighten up,' she said. 'Butch and Roxi will live outside from today. I understand your reasons. Really I do. It's just you've been so serious about them. You've hardly laughed since they arrived.'

'Making out I've slept with a minipig is hardly going to make my sides split,' I said in protest, and then chuckled despite myself. 'OK, you got me. If I hadn't insisted on kicking them out last night, this would've been the final straw for sure. Now please remove the livestock from my bed and allow me to get dressed in private.'

A little later in the day, under an early springtime sun, a small domestic operation commenced. With the French windows open, I was first to step out onto the patio. I carried the minipig ark in my hands with some sense of ceremony. This was a special moment, after all. Emma followed behind. She cradled Butch in her arms. Next, Lou and the little ones appeared behind her. Lou was clutching a little bucket of pig feed, which she shook to encourage the final member of this procession to leave the house. Roxi trotted out enthusiastically, clearly relishing the new scents and odours that met her snout. Hovering at the doors, holding Miso protectively, May watched the rest of her family as we departed for the foot of the garden. There, I returned the ark to the spot under the dogwood spray. The one I had chosen for it in the first place. With Butch and Roxi inside the enclosure, and the gate shut, I thought they looked just fine. It was my wife who voiced her reservations.

'I feel like we're abandoning them,' she said. 'It's the piggy equivalent of taking kittens to the river in a bag weighed down with stones.'

'Emma,' I said, reasoning with her. 'Nobody is suggesting we drown the minipigs or take them into the city and lose them. You might think their place is on your lap, but the truth is we're giving them the very best start in life. Just look around you,' I added, gesturing with my hand. 'They have grass here, fresh air and even the sun is shining. I'll open up the old chicken run so they can play about in there. They couldn't ask for more.'

At my feet, both minipigs were showing a real interest in the ground. I could only imagine how that must smell to little animals that had been kept on carpet and tiles for the last few weeks. Just then, Butch thrust his snout into the grass and pushed hard. It was like watching a living shovel, such was the ease with which he chiselled up a little strip and rolled it out of the way.

'I suppose he does look happy,' noted Emma, who crouched to replace the strip and pat it down. If Butch showed a talent for lifting the grass, Roxi proved to be more than a match. The way she ploughed into the soil straight after Emma had replaced it demonstrated surprisingly powerful muscles in her neck and chest. 'They're like kids with a new toy,' she added, raising her hands in submission. 'I guess they'll settle down soon.'

'They can still have visiting rights,' I said brightly. 'Under supervision, Butch and Roxi will be welcome any time, especially weekends, though I think you'll agree that overnight stays are out of bounds. The only other condition is that they don't place a trotter inside my office again. I need to feel that I can answer the phone without praying that the person on the other end isn't calling to further my career.'

Emma smiled, and linked her arm with mine. The sun was shining through the branches of the oak, dappling the light. Wood doves could be heard cooing nearby, while the little ones squabbled over where best to place the food and water bowls. As Lou made sure there was enough straw in the minipigs' ark, I just knew I had made the right decision. What mattered was that I stayed strong, no matter how much I would be tested.

The first twenty-four hours were tough. Not for me. I was overjoyed to have my spot on the sofa unchallenged. That afternoon I read the paper without risk of a minipig smashing through the middle. The rest of the family were notably quiet, and even a little tetchy with one another, but it wasn't until the evening that the reason became apparent. First the little ones played up to the absence of Butch and Roxi in the house. When they first moved in with us, Frank and Honey had quickly built into the bedtime ritual a need to pop into my office and say goodnight to them. Now the minipigs were outside, I had to be firm. Having brushed teeth and read several stories, the prospect of trailing down the garden was, in my view, a delaying tactic. Darkness was settling by the time we'd finally got them to sleep. Then it was my wife's turn to further hold up any chance of peace and quiet.

'Come and sit down,' I said. 'I'm sure they're just fine.'

At the window, Emma continued to peer through the curtains. 'They might be frightened. What if they get scared?'

I waited for her to face me. 'I really don't think pigs are scared of the dark. Even minipigs.'

Emma looked less than satisfied by my response. 'At the very least, we should've fitted a door to the ark.'

'But they pop out to wee in the night, as we know to our cost.' I turned her attention to the corner of the room behind the television. The carpet had been scrubbed so many times it looked worn and washed out. 'Emma, it's totally understandable that you're anxious, but you just have to tell yourself all will be well. Don't you remember what it was like when our children were babies? Each time we moved their cot out of our bedroom for the first time, neither of us slept a wink.'

'I remember,' she said, and returned to her station at the curtains. 'But there was no danger that any of our kids would be preyed upon by owls and foxes.'

'That isn't going to happen. Trust me.'

'Wait!' Emma cupped her hands against the glass and peered into the gloom. 'There's something out there!'

'Where?' The note of alarm in her voice persuaded me to join her at the window. 'What do you see?'

Emma pointed towards the rear corner of the enclosure, furthest from the shed.

'What if it's a badger?' She sounded panic-stricken. 'Those beasts are bloodthirsty. They'll tear those poor things limb from limb!'

Unquestionably, something was moving in the nocturnal shadows out there. Even with the moon behind cloud cover, I could see signs of activity; a dark shape that seemed to be tussling with something in the ground.

'I think you're confusing badgers with demons,' I said, as the creature in question came away with what must've been a buried root. 'Anyway, that's Roxi.'

Emma looked at me, and then back at the enclosure. 'Shouldn't she be in bed?'

'Not mine,' I replied, and suggested we just leave her to it. 'Your minipigs are fine. They'll still be there in the morning.'

At first light, on drawing open the bedroom curtains, my initial thought was that Roxi hadn't slept at all. Then I wondered whether we'd been the victims of some angry drunk in a stolen digger.

'Emma,' I said, staring in stunned amazement. 'There's something you need to see. Really. Wake up!'

After a moment of stirring and much muttering, my wife joined me at the window. She yawned first, covering her mouth with the back of her hand. Then she stretched her arms out wide, finding focus at the same time. That's when she gasped in a way that could've led to a shriek. Instead, she swore out loud.

'What's happened?' she asked, staring in disbelief at the enclosure. 'They've turned it into a *pigsty*!'

Outside, the section of garden now reserved for the minipigs looked like it had been rotovated by vandals. Sometimes, the chickens had left the grass a little worn in places. Overnight,

Butch and Roxi had ploughed up every last blade. Even the dogwood had been stripped back to stumps. Evidently their work had been exhausting. I could see them both inside their ark, sleeping soundlessly.

'I suppose it could be worse,' I said. 'At least it isn't winter time.'

'Will it grow back?'

'Don't know,' I said with a shrug. 'It's what pigs do.'

'Even minipigs?' Emma turned to me. 'Surely not?'

'They might be little,' I said. 'But they're still pigs at heart.'

Emma considered this for a moment. She seemed a little crestfallen, as if somehow I was spelling out something she had been avoiding for some time.

'I'll talk to them,' she said, before shutting herself away in the shower.

Over time, a pig will root up all the grass available to it. The damage doesn't just stop when the last blade has gone. As I learned before the spring was out, a pig heads deeper down.

Once I had overcome the shock of seeing the end of our garden in such a state, I began to pay attention to the behaviour of the perpetrators. Roxi really was the worst offender. Once she got a smell of an acorn in her snout, even if it had been buried for ages, she would not stop until she found it. She possessed a single-minded determination I have only ever witnessed in human beings between the ages of two and six. Every now and then I would take myself to the window to see how they were doing. Increasingly, all I would see of our female minipig was her hindquarters facing the sky. She didn't dig like a dog. Her front trotters were simply planted in the base of her excavations as a kind of anchor. It was her snout that did the hard work, throwing up clods of clay and stone.

With every hole she dug, the soil would simply get displaced into a previous effort, which gave her somewhere else to investigate once she had found her prize. For hours she would push, shove, turn and grunt until she found what she was looking for. It wasn't just ancient acorns. She unearthed an old horseshoe, bits of pottery and broken bricks that had presumably been discarded by the builders during the construction of our house. The only area of the enclosure that remained intact was that covered by the flagstones. There was no way that she or Butch could penetrate such a layer. Given what was at rest underneath, this came as a great relief.

Instead, they undermined it.

'There's something in with the minipigs,' said Emma one morning, having returned from feeding them breakfast. 'It's an oily rag, I think.'

I had grown used to reports of fresh finds. Some cast-off from a garage didn't seem worth setting down my toast to investigate straight away. So, I left it in favour of a morning's work. This was something I had come to really appreciate since moving the minipigs outside. Despite the chaos they had gone on to create in their own space, I could at least be assured of solitude and order in my office once more. In some ways, in the wake of what I'd been through, it didn't even feel like work.

Around lunchtime, I decided to stretch my legs by ambling down to see the minipigs. They had grown used to the sound of the back door opening, and were waiting eagerly to see me. I carried a watering can with me, which I used to fill their drinking bowl, plus two pears from the box we kept on the woodpile.

The pears I tossed over the fence for them. The watering can I dropped on the grass in front of their gate when I saw what it was that they had disturbed. It was arguably the most ghastly sight I had ever come across. When the phone started ringing in my office, I just stood there. Even when Butch and Roxi tried their hardest to alert me to the incoming call, a moment passed before I gathered my senses and raced back to my office. As it turned out, I should have just ignored the call. Instead, by the time I hung up, the prospect of picking up cat bones from the enclosure was the least of my worries.

13

This Little Piggy Went to Market

Emma bumped into Tom some days later. He'd asked her to remind me of the forthcoming livestock market. Emma had said that I'd love to join him. I don't know if she had explained that we were struggling. Either way, as soon as Tom enquired after the minipigs, on picking me up in his Land Rover, he saw straight through my positive response.

'When you say "brilliant",' he enquired, drawing up at the junction of the main road at the far end of our lane, 'do you mean "bloody awful"?'

I waited for him to pull out before telling him what had happened. It meant I wouldn't have to look him in the eye.

'Everything was going fine until they excavated Misty,' I said. 'As if that poor cat hadn't been through enough.'

'Is there anything left of her?' he asked.

I told him I had rescued most of the bones, even the chewed and splintered ones, and stacked them in the shed. Tom made no comment. We just motored on, stock box in tow, leaving the village behind.

'I did think about reburying her under the flagstones,' I said eventually. 'The problem is I very much doubt we would pass the inspection.'

This time Tom glanced across at me. Judging by the way he smiled to himself before returning his attention to the road, I figured Tom knew just what I was talking about.

'So,' he said, 'Trading Standards have tracked you down, right?'

I nodded solemnly. 'I had no idea that as a registered pig-keeper my details would be passed on to them. I told the lady she must have the wrong number, but she was insistent. She said she was calling to make an appointment to check we were complying with the law.'

'You come under the same rules and regulations as meat-producers now.' Tom sounded unsurprised by what I had told him. 'It doesn't matter that your minipigs are pets. They're all the same to the Trading Standards people.'

'But this woman,' I said. 'She made me feel like I had plans to open a hot dog stand or something.'

'She's just doing her job,' Tom cut in. 'They have a legal obligation to check you out and make sure the pigs are properly cared for.'

'She's arranged to pay a visit in a couple of weeks from now,' I said. 'The trouble is I've no idea if we're doing everything right. Only yesterday I caught the little ones taking a packet of cornflakes from the cupboard to feed them.'

Tom tutted. 'You could go down for that.'

'Yes, I'm well aware that it's illegal to give the minipigs a proper breakfast from the kitchen! Try telling that to small children!' Resting my head against the back of the seat, I breathed out long and hard. 'I'm sorry,' I said. 'It's just Butch and Roxi are still pushing all my buttons. Getting them out of the house was great, but now they've created a whole new set of problems.'

'How's the turf?' asked Tom cheerfully.

'Very funny,' I said. 'Please don't say you told me so.'

Turning off the main road, we travelled along a lane riddled with potholes and flanked by ragged hedges.

'They're only little,' Tom said eventually. 'How much trouble can they be?'

'It starts at daybreak,' I told him. 'With the dawn chorus. Only I can't hear the birds any more because the minipigs make such a racket. Honestly, they get louder and louder until they're fed.'

'They're hungry animals,' Tom said. 'But it's important you ignore the noise and feed them when you're ready. Otherwise they'll know that squealing at the top of their voices will always bring them food. You wouldn't pander to your kids in that way. The same rule applies to Butch and Roxi.'

'But my kids didn't make such a din that they threatened to disturb the neighbours,' I pointed out. 'We can't just let them carry on kicking off like this. We'll end up with an ASBO.'

'You have to break them out of the habit,' warned Tom. 'The longer it goes on the harder that will be.'

We drove in silence for a while. Eventually, Tom turned on the radio. He was right, though. I also knew it wasn't only Roddie I needed to be concerned about. At full volume, the minipigs were quite possibly a familiar sound to every last resident of the village.

'The noise is one thing,' I said eventually, having dwelled on the issue for too long. 'The flies are another.'

'Where there's livestock, there are flies, especially now it's warmer. You'll get used to it.'

'I don't want to get used to it,' I said. 'There's a little black cloud hovering at the foot of my garden. I'm worried it looks like we've had a curse placed over us. Can't I spray them or something?'

Tom chuckled to himself. 'Your flies are attracted by the

dung. Most owners just let their pigs trample it into the soil.'

'Most owners have land,' I pointed out. 'They don't keep their pigs in a garden.'

'Fair enough, but it's a great fertiliser. All I can suggest is that if you want less flies then you'll have to start clearing it away.'

'How? The bin men won't take it. I should imagine that contravenes a whole chapter in their book. It's just one more shortcut to prison.'

'Sell it,' Tom replied.

I looked across at him. 'Since when did pig poo come with a price tag?'

'Their dung is brilliant as compost. You can reduce your fly population *and* make a couple of quid out of it. As for your Trading Standards visit, I really wouldn't worry. It's like a VAT inspection with wellies. They can seem a bit fierce, but so long as they don't find a stack of licked-clean dinner plates in the pig ark I'm sure you'll be just fine.'

I had never visited a livestock market before. Nor had I even given one a second thought. As we climbed out of the Land Rover and made our way towards the entrance, it felt like I was approaching the turnstiles to another world,

'Everyone is wearing flat caps,' I said to Tom. 'I couldn't look more like an outsider if I tried.'

'I've never worn one in my life,' he replied, striding with some purpose.

'But people are looking at me funny,' I told him under my breath. 'Are you sure it's OK for me to be here? It's teeming with farmers.'

'*You're* a farmer,' he reminded me. 'On paper, at least.'

'Exactly,' I hissed. 'You only have to look at my hands to

see that I don't make a living off the land. What are the chances I'm the only one here who doesn't carry a shotgun licence?'

Tom was first to pass through the turnstile. For a moment I wondered if I should just wait in the car park until he had finished. Sensing someone much bigger than me breathing down my neck, who reeked of wood smoke and antiseptic, I hurried to join my friend.

'Will you relax?' Tom asked, as we faced the market. 'This isn't a Klan rally. Anyone with an interest in livestock is welcome. It doesn't matter what you do for a living. It's what you know that counts.'

'Which is very little.'

'And the reason why I thought bringing you here would be an education. Let's pick up a sales catalogue and you can help me choose my next pigs.'

'What are you looking for?' I asked, following him to a sales stand.

'I have to say I'm tempted by some gilts.'

I looked at Tom bemused. 'I realise you're not talking about financial investments,' I said. 'But you'll have to help me out here.'

'A gilt is a female pig,' he told me. 'But I'll probably go for the boar once more. Gilts are just fine, but they can be a bit moody when they're brimming.'

We were standing on a walkway inside a large agricultural courtyard. The space was divided into pens housing different farm animals and breeds. Each pen was surrounded by ruddy-faced men wearing the compulsory caps. Tom went without headwear, but then he had a manner that suggested he could produce one if required. It was noisy, not just with the hum of chatter but the occasional bellow from a bull. It felt like I was at a cross between a festival and a farmyard.

'Brimming,' I said eventually, having pondered in vain what Tom meant by that. 'Should I have a guess?'

Tom was leafing through a catalogue he had just picked up from the stand.

'No need,' he said, and invited me to follow him through the crowd. 'Let's make this a hands-on experience.'

Most of the pigs were to be found along the back row. If I had thought that Tom's last herd were big, some of the specimens on show here were enormous. Others lacked the size, but that was countered by their sheer ugliness.

'Man,' I said, grimacing at one specimen. 'There's a face that only a mother could love.'

'She's a beauty,' Tom replied, stopping before the pen. 'And it would be rude not to check to see if she's brimming.'

'With what?' I asked, as he rolled up his sleeves. 'Excitement?'

'You could say that.' Reaching into the pen, Tom placed both hands upon her rear haunches. 'When I push, let's see what happens.'

Silently, I watched him place a little weight on the back of the pig. She didn't like it one bit, and moved off with a grunt.

'What does that mean?' I asked. 'Is that good?'

'Not if you're planning on breeding her three weeks from now.' Tom looked around, scoping out the pigs. 'A gilt that's brimming is basically in season. It happens regular as clockwork, and lasts for three days. With pigs, in fact, *everything* happens in threes.' He stopped there, and began to count off on three fingers. 'She's fertile for three days, every three weeks, right? Even the gestation period runs along the same lines. Three months, three weeks and three days after a pig conceives, you can expect a litter of piglets.'

'Wow,' I said, impressed. 'That sounds more like sorcery than nature.'

'Once you know your window of opportunity,' he continued, ignoring my comment, 'you can put a breeding plan in place. Here, try this one yourself. Let's find out if she's brimming.' Crossing the walkway, Tom invited me to check out a black pig with ears that hung over her eyes like curtains. She had her nose in a trough and her back turned to us. 'By pushing down on each side,' he said, as I lined up to follow his instruction, 'you're basically mimicking the action of a boar mounting her.'

I hesitated as he said this. 'I feel a headache coming on,' I told him, but pushed down nonetheless.

This time, the pig did not protest. Instead, rising to meet my palms, she shuffled and spread her hind legs a little. Tom smiled.

'She's ready for you to serve her now,' he said. 'That's pig-speak for—'

Before he could finish, I cut him off by lifting my hands as if in surrender. The pig turned and shook herself down, looking somewhat disappointed.

'So, if that had been a boar on her back, would he have done the business straight away?'

'Absolutely.' Tom was still watching the pig, as were several other farmers who had gathered to inspect it. 'They don't have time for foreplay,' he added. 'You need a boar that doesn't mess about. Otherwise, the gilt will lose interest. That's why it's so important to find a boy who's in top condition.'

Facing Tom, I just knew where this was heading. 'Please don't make me do that,' I said. 'Can't I just watch you?'

Tom considered my request, rolling up his catalogue at the same time, only to shake his head.

'By the time this auction starts,' he said, 'you'll have a solid feel for the pigs.'

* * *

Had I known that I would find myself behind a feeding boar that morning, preparing to weigh his testicles in my hands, I'd have packed a pair of surgical gloves. Even if I'd produced them from my pocket, however, I imagine Tom would've confiscated them. As uncomfortable as I felt, I recognised that my friend was giving me the chance to get to grips with pig keeping so that I might return home with peace of mind.

On my knees in the pen, with Tom watching over me, I cupped my palms together. I was looking at the back end of a medium-sized, rust-coloured boar, with his bollocks just inches from my face. Merely thinking about the task I faced left me feeling a little bit sick.

'Are you sure he's not going to kick me in the teeth?' I asked. 'If it was me, I wouldn't react well.'

'Relax,' said Tom. 'While there's food in the trough, nothing else matters to him.'

I turned back to the boar, and focused on the swollen sack between his legs.

'I don't like keeping secrets,' I said, 'but can we never speak of this again?'

Very gently, I closed my hands around the appendage. I was surprised by how warm they felt.

'Go on,' said Tom, encouragingly. 'Give them a good feel. He won't mind one bit. He'll probably enjoy it.'

'At least one of us will.'

Having overcome the shock of connecting with the pig's crown jewels, I found my confidence rising. Mindful not to squeeze too hard, I began kneading his testicles as carefully as I could. They felt remarkably firm, like two grapefruit that had found themselves inside a hot-water bottle. By now, several farmers had gathered beside Tom to watch proceedings. I felt somehow that by doing this I had climbed to a level of understanding they could only admire.

'Of course,' observed Tom, leaning on the railings now, 'a seasoned pig breeder can tell if he's in good health just by looking.'

I snapped my hands away as soon as he said this. Tom beamed at me. Beside him, the farmers were looking at me warily. I climbed to my feet, trying to look confident, and brushed sawdust from my knees.

'I think this one is fine,' I said, dusting down my palms. 'A healthy specimen.'

'Really?' said one farmer. 'He looks clapped out to me. I wouldn't let that saggy-backed old bugger climb on one of my girls.'

The other farmer nodded and gestured at the pig. 'There's a lot of fat on him instead of muscle,' he pointed out. 'You really didn't need to cup his nuts to see that for yourself. Unless you prefer that method, of course.'

Still grinning, watching me blush, Tom explained that he was just teaching a newbie a few things. With my cheeks burning, I climbed out of the pen and wondered if we could go home yet. I noticed that most people had begun to drift towards the other side of the courtyard. When a figure clambered up onto the planks that formed a kind of catwalk between the pens, I realised to my relief that the auction was starting.

'Has anything caught your eye?' I asked Tom, as the two farmers left us alone.

'Not yet,' he said with a shrug. 'But we'll pass on this one, eh?'

Together, we headed towards the crowd. The auctioneer was standing over two cows, babbling away in what sounded like another language. He barely paused for breath, and yet somehow managed to acknowledge every bid with a nod of his head. I watched from the back for a good half hour. Observing the process was mesmerising as the auctioneer moved on to sheep and then the pigs. This is where it got interesting. Not only was I looking at each pen of livestock with a more informed eye than ever before, I noted Tom putting in the odd bid every now and then. It was only when we reached a pen containing four white, bat-faced piglets, however, that he began raising his hand with real determination. I had become so gripped that when my mobile bleated I killed the call straight away. When it rang again, I noted the name displayed on the screen, and realised this was someone who would never give up.

'Emma,' I whispered into the phone, as Tom traded bids with just one remaining farmer in the auction. 'Is it urgent?'

'I've just found Butch and Roxi in the garden,' I heard her say. *'The gate to the sty was wide open.'*

I asked her to repeat herself, just to be sure that I heard her right. It was a good thing the minipigs hadn't ventured out into the lane. My concern, however, was for their reason for hanging around on the lawn, which could only be very bad. Thinking through the implications of Emma's news, I pressed my palm to my forehead.

I didn't hear the auctioneer acknowledge my bid. I just saw Tom look at me in astonishment, before raising me by another fifty pounds.

When the other farmer declared himself out, I found the auctioneer looking directly at me. Aware that I would already be footing the difference, I didn't dare move a muscle until the hammer fell.

14

Under Inspection

The minipigs saw a lot of me in the run-up to the visit from the Trading Standards lady. Whenever I took a break from working, I would head down to the pigsty and make sure that everything was in order. I kept their water bowl full, ensured the straw in their ark was fresh, and the flagstones level. I didn't like to see them wonky. It just reminded me that they once marked the grave of the family cat. I even took a tip from Tom and created a wallow. There was no need to dig a hole before soaking it with water. Butch and Roxi had done that for me. It was the perfect place for them to stay cool. When the sun beat down they would loll about in their shallow pool and watch me make my final preparations.

It meant that when the doorbell rang at the appointed time, I was ready.

'Hi there,' I said brightly, addressing the woman with the clipboard pressed to her chest. She was wearing a tweed coat, buttoned to the neck despite the fine weather, and a pair of angular spectacles that did little to soften her manner. She didn't smile. She just offered me the kind of handshake that made me think of a dead fish.

'I'm here to inspect the pigs,' she announced.

'*Mini*pigs,' I said, hoping to impress her. 'They're not as tiny as when they first arrived, but I think you'll be surprised.'

'The media have a lot to answer for,' she said disapprovingly. 'No doubt you won't be the last minipig owner I shall have to visit.'

I held onto my smile as ably as I could.

'Butch and Roxi have certainly been looking forward to meeting you,' I said, aware that her presence on the doorstep had stirred them. As she turned her attention towards the garden, I just knew what her first comment would be. I figured it was probably best to get this bit out of the way first.

'It seems they've had the run of your lawn,' she said. 'Do you let them out on a regular basis?'

I faced the garden, wondering what to say. It was criss-crossed by random furrows from one side to the other. To be fair, Emma had taken full responsibility for not shutting the gate to the pigsty properly. Even so, when I returned from the market to survey the damage caused by Butch and Roxi, I was taken aback by the extent of their spree. Chunks of turf were scattered everywhere. Flowers and bushes in the beds under the windows had been torn out at the root. It looked like the horticultural equivalent of a slasher movie.

'A nice lawn is not as important as the fitness of our minipigs,' I offered after a moment. 'A healthy pig is a happy pig. Isn't that what they say?'

The woman from Trading Standards placed what looked like a cross on the form pinned to her clipboard. I couldn't be sure if that was a good or bad thing.

'Generally, I inspect farms and slaughterhouses,' she said, and faced me once more. 'This will be my first house visit.'

'Won't you come in?' I suggested, aware that she was keen to get on.

As I invited her to follow me into the kitchen, I hoped she would note the aroma of coffee on the hob and freshly-baked bread which I'd just rushed back from the supermarket. She wasn't here to view the property, of course, but I wanted to make the very best impression that I could. As far as I was concerned, the prospect of falling foul of trading standards was unthinkable. The sheer embarrassment if word got out around the village would be too much to bear.

'Mr Whyman, I couldn't help noticing a sign outside your house that concerns me.' Declining my offer of a coffee or tea, she took a seat at the kitchen table and set down her clipboard. 'You're advertising free poo.'

'That's right.' I sat across from her. 'I'm all for recycling. It's a civic responsibility, in my view.'

I had only put the board out this morning, along with three refuse bags filled to the top and tied with string. The lady from Trading Standards regarded me through her glasses. Every time she blinked her lashes brushed the lenses. I knew for sure that I hadn't contravened any laws. I wasn't stupid. I'd checked out the situation online. What I had to offer was officially classed as a fertiliser, which meant I could sell it. I wasn't sure how connected she was with the tax man, however, which was why I had decided not to charge money for each bag. At least not until she'd ticked us off her list.

'What kind of poo is in those bags?' she asked, and tapped her pen against her clipboard.

'Pig poo,' I said, a little surprised that she had to ask. 'What else would it be?'

'Then can I suggest you state that clearly on the sign.' She caught my eye and smirked. 'We wouldn't want people thinking you were hoping to give away human faeces here.'

I wanted to ask her what on earth she thought I was like? Of course people wouldn't think that. Unless they were council busybodies with nothing better to do than check we weren't planning on turning our pets into pie filling.

Instead, I assured her I would take care of it. 'Now,' I added, anxious to move on. 'Where do we start?'

Without lightening up by even a watt of charm, she asked to see my paperwork. I had everything ready for her in a folder. It contained all the correct farming credentials. My hope was that she would realise that I took my responsibilities seriously, despite being the registered owner of two silly toy pigs.

'This is fine,' she said eventually, having studied the documents so closely I thought she was sniffing them. 'Of course, you'll need to worm them both regularly. You do know how to do that, don't you?'

'Of course,' I lied, and made a mental note to ask Tom about that.

She closed the folder and then wrote something down on her clipboard. 'What method do you favour?'

She was still writing when she asked me this. I swallowed uncomfortably.

'The back end?' I said, thinking some kind of worming suppository had to be available. She stopped writing for a second, finished by adding a big cross and then looked directly at me. 'What do *you* prefer?' I asked.

'I'm unfamiliar with anything other than injection or feed form. I wouldn't recommend anything else.

'Understood,' I said. 'Feed form sounds just fine.'

'It is if you're worming a herd of fifty pigs or more,' she replied, as if testing me somehow. 'Generally, you can only buy feed wormer in volume. I would suggest you inject your minipigs at the appropriate time.' She rose to her feet, gathering her clipboard at the same time. 'I'll just collect my boots and things from my car, and then perhaps we can visit the pigsty?'

I agreed that boots were a good idea, giving silent thanks that she hadn't walked into a house in which Butch and Roxi took priority seating.

'Why don't I meet you down there?' I suggested, keen to make a quick final check and guarantee nothing but ticks on her clipboard.

As soon as Butch and Roxi heard my approach, they barrelled from their ark to greet me. At least that's how Emma always viewed it. As far as I was concerned, they were simply crowding the gate in the hope that I would feed them.

'Behave yourselves,' I said quietly, and opened the latch. 'Come on now. Move out of the way and let me in.'

Thankfully, we had no legal obligation to train them. Emma still had high hopes that one day her minipigs would be able to tie shoelaces or whatever it was she had in mind. As things stood, I wasn't even sure they responded to their own names. All I could do was muscle them out of the way having opened the gate. When they first arrived, we had to be careful not to tread on them. Now, it required some effort to shove them gently to one side with one foot. I glanced over my shoulder. The lady from Trading Standards had yet to return from her car. It meant I could throw a handful of pig feed into their trough. Anything to keep them from hassling her when she finally showed up.

As they dived on the food, I scanned the pigsty. Everything looked just fine. I hadn't missed any cat remains in my clearing up operation. Nor had the minipigs collected a colony of cockroaches or anything else that might contravene laws. With no consumer durables within the pigsty for Butch to horde away, he had focused his thieving instinct upon twigs that dropped from the tree. I double-checked the ark, just to be sure he wasn't fencing car radios for the local squirrels or anything like that. I was pleased to see that it contained nothing but straw. When I heard the garden gate open up, I turned in readiness to greet our guest, only to be left a little lost for words.

'Sorry about the wait,' she said, on waddling across the lawn. 'It can take a while to get all the gear on.'

The lady from Trading Standards hadn't just changed her shoes for some wellington boots. She had also hauled on what looked to me like a biohazard suit. She still had her clipboard with her, but I waited until she turned to close the gate to be sure she hadn't also got some kind of helmet or breathing apparatus hooked to the back of her utility belt.

'That's quite an outfit,' I said, as she joined me. 'I feel a little underdressed now.'

She cast her eye at the minipigs. It was the first time I had introduced them to someone who didn't at least comment on their cuteness.

'There's nothing worse than returning to the office caked in mud and slurry,' she explained, and began making yet more notes. 'I make no exception.'

'I see,' I said, quietly hoping that nobody I knew would choose this moment to walk down the lane past our house. The way she was dressed it looked like we were host to some kind of awful disease outbreak. 'Anyway,' I continued, keen to get this finished, 'as you can see Butch and Roxi have everything they need to be fit, happy and healthy minipigs.'

The lady from Trading Standards navigated her way across the mounds and trenches they had created in the soil. She bent down before the ark and peered inside.

'I suspect before too long you might need larger quarters,'

she advised me. 'They're still growing pigs, after all. Especially the female by the look of her.'

By now, Butch and Roxi had cleared their trough. Both were at my feet, nudging me and whining for more food.

'You really don't have to worry about that,' I said, and gently scooped Butch away with my foot while she wasn't looking. 'I very much doubt they'll get much larger.'

She seemed to smirk when I said this. I couldn't be sure because I was straining instead to see what she was scribbling on her notes. It was impossible to tell, but I did catch her finishing with a tick before moving to the other end of the pigsty. I gave her some space. Despite the fact that Roxi had returned to her favourite pastime, which was basically the destruction of the earth's surface, I figured we were almost home and dry.

'What's under here?' she asked, having moved on to stand on the flagstones.

'*Nothing*!' I said, a little too quickly. 'Nothing at all.'

I had done my level best to fill the cavity left behind when the minipigs plundered Misty's grave. I didn't think it would give way as she tested it now. I was just worried by all the attention she was giving it.

'You'll need more hardstanding than this,' she said eventually. I would recommend a concrete area beside the gate.'

'Concrete? In my garden.'

'You've been lucky,' she said, and stepped off at last. 'We've had dry weather for a long while, but as soon as it rains, this area is going to get very bogged down indeed. For pigs this size, even if they have got more growing to do, it wouldn't be fair to them.'

'Of course,' I said hollowly. 'I'll get onto it.'

'Very good.' She placed another tick on her clipboard, but

this one did little to raise my spirits. 'Can I also recommend that you have the boar castrated,' she added, without looking up.

'Why's that?' I asked.

'See for yourself,' she said, and gestured at the minipigs.

I had been so distracted by the flagstones issue that I hadn't paid any attention to what Butch was up to. At first glance, I thought he was simply trying to get a better look at what his sister was attempting to dig up. It took a moment for me to realise that for whatever twisted reason he was attempting to mount her.

'Stop that!' Immediately, I reached down and hauled him off. 'They're just playing!' I added, trying not to sound as flustered as I felt. 'Kids, eh?'

'Maybe they are,' she said, 'but it won't be long before the testosterone seriously kicks in. Mark my words, when that happens there's no way you'll be able to stop him. I recommend testicle removal at the earliest opportunity. You really don't want to start inbreeding, Mr Whyman.'

'Absolutely,' I said, as if somehow that was even something I would contemplate. Still reeling from what had just happened, and aware that she was scrutinising me heavily, I filled the silence by adding, 'We don't tolerate incest in this household.'

I couldn't see whether she completed the form with a final tick or a cross. I was too busy focusing on my boots, wondering if it was too late to clarify what I had just said.

From the moment we rehoused the minipigs outside, I would awake each morning in a state of alarm. The weeks that followed our inspection were a case in point. At sunrise, Roxi would crash from her ark as if fleeing from a bad dream. She would then proceed to squeal, shriek and bellow for her breakfast. To give you an idea of just how hideous that sounded, try drawing the breath into your lungs as if it were your last, speed up the sound fourfold, and play the recording through a megaphone. In pig-speak, it would translate as something like, '*hey, we're dying out here! Horribly! And you're too bone idle to save us!*' Unlike a dog, which pretty much barks at the same volume, Roxi seemed quite able to crank herself up with no apparent upper limit. Every now and then, she would raise the volume somewhere close to triggering car alarms. It was as if to broadcast the fact that unless someone came down immediately and filled her trough then windows would be shattered.

Her methods didn't trouble Emma quite so much. Somehow, she could doze through it whereas I would be bolt upright in bed just as soon as the minipig made her presence known.

'Shall I feed her?' I offered anxiously one time, as birds took wing from their roost in the oak. 'I really think one of us should go.'

'Butch and Roxi are my responsibility outside of working hours,' she said, stirring at last. 'And I'll feed them when she realises this is not the way to summon me.'

'Are you sure it's you she's trying to summon?' I asked, raising my voice to be heard. 'Not the Devil and his minions?'

'She's just young. Any baby cries for food, but they grow up eventually.'

'At the very least, it's loud enough to be classified as a nuisance.'

'Nobody's complained.' At last, Emma hauled herself out of bed and climbed into her dressing gown. 'Apart from you.'

'It's reached the point where sometimes I wake up before it all kicks off. Then I can't get to sleep again.'

Emma faced me from the door. 'You were fine before this

thing about the concrete. I don't want the minipigs to become an issue, Matt. It isn't their fault that Trading Standards are making us jump through hoops.'

I pressed my fingertips to my eyes. All this talk before a cup of tea was giving me a headache.

'Tom is coming to help me with the hardstanding at the weekend,' I told her. 'He's also offered to fix the damage to the lawn. Not that there's going to be much lawn left.'

'Maybe you can ask him to take a look at the ark.'

'What's wrong with it? The ark is fine.'

'I think you're kidding yourself,' replied Emma. 'They're getting tight for space in there.'

By now, Roxi was making such a racket that I had to ask Emma to repeat that last bit.

'I'll do whatever you say,' I told her, grimacing at the noise, 'if you'll just go sort out your pet pigs.'

15

On Loss

I admit that I was upset about the concrete. Somehow, it represented the fact that the minipigs had become a permanent feature. It wasn't like one of those giant trampolines that had sprung up in back gardens across the country. Those things were an eyesore and a health hazard, but once the kids had broken bones you could at least dismantle them. Butch and Roxi weren't dangerous in any shape or form, but they would be taking up my time and space for the rest of their natural lives.

The cost of keeping minipigs was also mounting. On top of materials, I would have to pay Tom to lay the hardstanding for me, as well as slip him the fifty pounds I'd managed to add to the auction cost of his piglets. Then there was the issue with the ark. Tom had built the first one in exchange for my help in rounding up his pigs. It wasn't exactly a fair deal. I suspected he had done it for the love of his craft, and out of curiosity. Despite Emma's concern, not to mention the lady from Trading Standards, I decided to delay pressing Tom into constructing larger accommodation. On top of everything else, I just didn't think we could afford it that month. As neither minipig complained about their cosy quarters, I figured they could live with it for a little while longer.

I wasn't in denial about the fact that Roxi was developing

at a swifter rate than her brother. Not only was Butch markedly less noisy, he still wasn't much bigger than Miso. His sister, on the other hand, was becoming really quite meaty, which put paid to my belief that she was basically a pair of lungs on four legs. At mealtimes, she could headbutt Butch away effortlessly in her bid to be first to the trough. He didn't seem to mind too much. Whenever I glimpsed them asleep in the straw, he was cuddled up happily against the pillow-sized belly of his chunkier sister.

As a measure of the minipigs' cuteness, I noticed Lou becoming increasingly reluctant to show off Roxi first whenever her friends made the pilgrimage to see them. She was small, compared to a standard pig, but Butch had the edge. I hated to say it, and so kept it to myself, but Roxi also lost out in terms of her looks. While Butch could've been a cross between a wild boar and a teddy bear, with a sausage-like body and stubby little legs, his sister appeared to be teetering on trotters shaped like cheap slingbacks. It didn't help that her thick ginger lashes were so long they could've been fake, and there was no disguising the fact that her sizeable ears and snout were out of proportion with the rest of her body. To sum it up, she looked totally different to her brother, and not in a good way.

Emma could see it as well, but rather than opt for favouritism, Roxi's physical misfortunes pricked at her need to care and nurture. That's why she never complained about the noise her less pretty minipig could create, or the loss of any chance we had of a lie in. The fact that we were beginning to trail dirt from the pigsty back to the house went without complaint from her. Before and after work, she would mop the floors both voluntarily and furiously, and refused to let it dim her spirits. As far as Emma was concerned, the minipigs were a part of us, and she would do whatever it took to look after them. It made her happy, in a pained kind of way, and I had no desire to change the situation. It was all about family for her, after all. I also knew from experience how shattering it could be to have that undermined. Having experienced the kind of stable and loving upbringing that Emma considered to be ideal, I'd gone on to watch my family fall apart.

My sister, Annie, was born exactly two years after me to the day, late in March 1971. My earliest memory, in fact, is of being dropped off in a hurry at my godmother's house on my birthday. I remember standing on a gravel drive, clutching her hand as my parents drove away promising 'a big surprise' on their return.

The following afternoon, in a bid to keep things nice and calm, they left baby Annie asleep in the car for a while when they came to pick me up. Finally, when my mum told me they had something special for me, we trooped outside and gathered at the back of the car. It was a model that sported wood-panelled double doors on the back. These, my dad opened with great reverence and introduced me to the newborn swaddled in her Moses basket.

Two years later, when my brother was born, my parents repeated the process. As a result, possibly until I was age nine or ten, I believed that all babies came from the back of Morris Travellers.

My sister and I weren't especially close as children. We got on just fine, but though we shared a birthday our characters and interests differed. Annie was a wild redhead who could swing from crippling shyness in public to startling strops at home. She struggled at school, largely on account of a hearing problem in one ear that went undiagnosed for many years, but was supported throughout by our parents. By the time I left home for university, she had found herself secretarial work; a job that afforded her freedom in the form of a car, and the chance to forge a relationship with a local boy she would go on to wed.

In early 2000, when she called to say they were expecting their first child, Emma and I were striving to raise two. Through her pregnancy, we didn't see anything of them, mostly because Annie and her husband had moved north to the Pennine town where my dad grew up. When we did speak she was thrilled with their new life, looking forward to becoming a mother, and settling in just nicely. They even took on a black Labrador puppy, which made me chuckle when I thought of the extra workload this would bring. It was one more reason why I was looking forward to seeing her.

When we did make the visit, however, following the birth of their baby daughter, it was with a sense of injustice, shock and uncertainty.

A week before her due date, a routine scan had picked up unusual shadows. The doctors elected to perform a caesarean, delivering a healthy girl into the arms of my brother-in-law, along with the news that Annie had been diagnosed with ovarian cancer. She was 28, a first-time mother, with a serious illness that nobody could quite believe. Immediately, my mother resigned from her job to take care of her daughter and young family. Refusing to believe that the worst could happen, she dedicated herself to supporting Annie, her husband and baby, and stayed with them for long periods. My dad was unshakeable in his belief that Annie would get through this, and yet seemed lost whenever he was left alone.

I'll never know why my mother kept quiet about the fact that she was in pain herself. I can only think she just didn't want to add to the distress our family was in. Eventually, it became evident that she was living with something far more serious than occasional heartburn. On receiving the call from my dad, confirming that she had been diagnosed with cancer of the oesophagus, I remember feeling as if we were being toyed with by some cruel but unseen power. When I broke the news to Emma, we just sat facing each other in disbelief.

By now, my sister was undergoing intensive treatment, including chemotherapy and radiotherapy. My mum quickly joined her, and in many ways began to mirror her prognosis. They both knew they were very ill, but clung to the diminishing chances their doctors offered. Watching Annie's baby girl develop was a source of great joy and immense heartbreak for them both, as it was for us all. Towards the end of that year, my sister began making plans for her daughter's future. She wrote her a series of long letters, which took the shape of a book, promising to look over her at all times. Above all, Annie was terrified that she would be forgotten, despite assurances from her husband. He showed great strength throughout this time, living as he was with the unthinkable.

By Christmas, in great pain after countless operations, Annie made the decision to end her treatment. She wanted to spend what little time she had left with her family, and so a bed was set up in her drawing room, with a view across her garden and the charcoal elevation of the moors beyond. Whenever Emma and I visited she would delight in showing me how my niece had grown in a year, before drifting into tears and then sleep. As January progressed, she retracted her hopes for the future into the belief she would live long enough to see the snowdrops grow in her garden. Quietly, we could see that wasn't going to happen, only for her Macmillan nurse to show a small act of great kindness by travelling out to where they had flowered, and picking some for a little vase on her ledge.

Annie's final wish, two months before her thirtieth birthday, was that her family should be with her when she passed away.

It was my dad who phoned me to say the time had come. For days I had been braced for the call, and when it arrived I simply went into autopilot. I said goodbye to Emma, and waited for my brother to collect me in his car. Together, in near silence, we travelled north. It was early evening by the time we crossed the moors. The journey was bleak on several levels. Some kind of controlled heather-burning operation was taking place. All around, we could see columns of smoke rise up to a certain level, and then spread across the sky. On making our way down the hillside road towards my sister's village, it felt as if we had arrived in another world.

My brother and I were the last to arrive. Our parents were present, along with a nurse, Annie's husband, their dog and little daughter. Over the course of an hour, we took turns to speak to her in private. My sister's bed occupied the centre of the room. She was heavily sedated but able to whisper the odd word and sometimes open her eyes. She looked very different from when I had last seen her, physically defeated in so many ways and with a focus that was just lost. I told her that I loved her, something I'd never said before. I also offered to honour our shared birthday with a present for my niece each year. It felt so strange to talk about a future without her, but she acknowledged my voice. As dusk settled, we retreated to the kitchen so she could share a moment with her beloved child whose bedtime beckoned. When my brother-in-law encouraged their daughter upstairs to her room, trying his best to keep things normal, we filed back in to sit at Annie's bedside. By now she had stopped opening her eyes, but the nurse assured us that she was listening.

I had never witnessed anyone pass away before. I'm not sure what I expected. I was just totally unprepared for such a slow fade into death. Each minute felt like an hour. Each hour consumed the next, long into the night. Sometimes we told stories, reflecting on her life. Once or twice, as one of us shared a memory a flicker of a smile would cross Annie's lips. Mostly, however, it seemed as if we were cocooned inside a house of silence.

Throughout, her breathing continued to shallow. Eventually, it became little more than a series of gasps, and yet each intake of air had some fight to it. When the last breath left her body, out of nowhere it seemed, I felt profoundly shocked. Despite knowing this moment would come, it just didn't seem real. My brother-in-law had a baby monitor stationed in the room. Moments after her passing, it crackled as my niece turned in her cot in the room above. In a way it was a reminder that life went on. Soon after, I remember taking my turn to kiss my sister on the forehead. As I did so, I knew that Annie had gone.

In the hour that followed, nobody knew quite what to do. There were procedural matters that needed to happen. The doctor was summoned to record Annie's death, and when the undertakers came they removed her body with trained respect but uncomfortable efficiency. Watching them load the coffin into the back of the hearse and close the rear doors, all I could think about was the moment I had first set eyes on my sister in my godmother's drive.

Soon after, I called Emma. It was the middle of the night. She picked up on the first ring. I told her what had happened. She cried. I had yet to shed a tear. I felt as if I couldn't do so in front of my family. It was not something I had ever done before. The same went for my brother and my dad. In the grip of her own terminal illness, I couldn't begin to think what emotions my mum was going through. We just fell in and out of sentences, stopping when it felt like our emotions might get the better of us. Throughout, I was keenly aware that this was my brother-in-law's house. We were grieving for a sister and a daughter, and yet it was here that he had spent his married life. At one point, he went to check on his little girl, only to return ashen faced. He took me to one side, and swore to me that at bedtime he had left her nursery in a mess.

'It's tidy now,' he told me, choked by what he believed this meant. 'All her toys have been put away.'

Before daybreak the next morning, on a camp bed in the box room, I woke up feeling suffocated. My only need was to be home with Emma and the kids. My brother wanted to stay, and so my dad offered to give me a lift to the station. In his own way, he had attempted to remain as resolute as possible, for his own sake as much as everyone else's. Everything I had witnessed felt like it was still ablaze in my head. Above all, I couldn't shake what my brother-in-law had told me. As we drove out towards the station, I felt that I should share it with my dad. He said nothing while I repeated the account. It was only as I reminded him how the baby monitor had crackled with static just seconds after Annie passed away that I realised the car had slowed to a crawl. I turned to see that he was in tears. I faced the front, confused as to whether I had done the right thing. A moment later, he cleared his throat and eased the car back up to speed.

It was a long train journey home. I'd packed an old Game Boy in my bag. I played it endlessly. I found that whenever I looked up at my fellow passengers, a catch would form in my throat. It was only when I finally made it home, into Emma's embrace that the lid came off my grief. In the week before Annie's funeral, I barely spoke without my voice cracking. At the service, standing with Emma, my parents, my brother and my sister's widower, I had this awful sense that this was only the beginning, not the end.

My mother's health deteriorated soon afterwards. It was as if she had invested all her strength in being there for Annie. Her final weeks were painful, sad, stricken by grief and a silent rage that all this could happen to one family. In our own individual ways, my dad, my brother and I each went on lockdown. We dealt with what was happening as only we knew how. A little over three months after we had gathered to say goodbye to my sister, Mum died alone, at night, in a hospital ward. My dad telephoned me with the news, having spent the preceding afternoon at her bedside. There, he had shown her the photographs taken the day before, on a trip to the London Eye for family and friends, which she had helped plan to mark his sixtieth birthday. It had been a subdued event. Everyone knew what was coming, and I didn't shed a single tear when I heard. Things had gone too far for that. At her funeral, in the same church, with a plot in the same row as Annie, I felt so numb I did not think I would feel anything ever again.

All I recall is that Emma stood beside me, my hand clasped in hers, and she did not let me go.

Ten years on, we had created the family she always wanted as a child. For me, it offered every one of us the kind of warmth and security I had taken for granted when I was younger. Having lost so much so quickly, I wasn't going to let two unusual pets get the better of us. I may not have been prepared to lay down my life for the dog, but I would do so for my wife and children without a second thought. Metaphorically,

I hoped to demonstrate that by laying down concrete instead. In the big scheme of things, once I'd thought things through, all my fussing and objections to having minipigs in our midst seemed somewhat trivial. There were more important things in this world, after all.

16

Muscle and Sweat

In a bid to control the cost of creating a hardstanding area for the minipigs, I had volunteered my services as a manual labourer. I'd phoned to tell Tom, and he had seemed more than happy to take me on. I just wondered whether he'd forgotten our conversation. When he arrived on the appointed morning, one sunny Saturday, I discovered he had brought his son to help him.

Jake took after his father. One look at him told you this wasn't a young man with a desk-bound job and a chronic social networking habit. In his early twenties, he was crop-haired, lean and tanned, as you might be had you just come back from a tour of duty with the Parachute Regiment. His solemn air was down to two things: a massive hangover and the fact that he'd been forced to take leave on medical grounds due to an injury to both ankles. I had visions of him plummeting to earth with his chute in ribbons as he came under heavy fire. Privately, Tom told me his boy had taken a tumble in training. I didn't like to raise the issue directly with Jake. Just in case it upset him and I ended up losing my limbs from their sockets. Instead, we left him to operate the grab hook on the flatbed lorry that had arrived. It was here to deliver an industrial mixer Tom had hired, along with two tonne bags of gravel and a stack of sand and cement.

Tom said it would take his son a couple of minutes, and suggested we check out the pigsty.

'Roxi's grown.' That was his first somewhat unimpressed response as we fought our way through the gate. 'Unlike my little boy here! How you doing, Butch? What's that? You want a tickle just there? You like that do you? Ooh, yes you do!'

'Tom . . . *Tom*!' I waited until I had his full attention, and then pointed at the ground. 'So, the lady from Trading Standards said we need to concrete a third of this.'

'That's the easy bit,' he said, giving Butch a final pat to his little flanks. 'First we need to dig it out. Nasty job. It takes muscle and sweat.'

'I'm up for it,' I told him.

Tom looked back towards the drive, a little anxiously, it seemed to me.

'I'm sure Jake can help once he's finished unloading. He needs to earn his keep, after all. I tell you what, though,' he added, turning back to me, 'two cups of tea would go down a storm.'

Before Tom and his son had shown up I'd told Emma and the kids that I would be outside working hard all morning. I very much doubted they would see much of me, I had warned them.

'I'll get the kettle on,' I said, and left Tom to temporarily fence in Butch and Roxi on one side of the pigsty.

I found Emma watching Jake from the office window. He had already fired up the cement mixer and was busy shovelling in hardcore from one of the open bags.

'Hi!' she said, sounding surprised to see me. 'How's it going?'

'Fine,' I told her. 'Thirsty work.'

Emma returned her attention to Tom's son. 'He's just asked to borrow your hose,' she said. 'I told him that was fine, though

I wasn't sure it was long enough to reach around from the tap in the yard.'

I looked out of the window. The hose was nowhere to be seen.

'There's plenty of length,' I assured her. 'Leave it to me.'

'Shall I fix the boys some drinks?' she asked as I made my way out.

'Builder's tea for us all,' I said. 'No sugar for me.'

With the cement mixer revolving at full tilt, Jake didn't notice me emerge from the yard. I stopped hauling at the hose. As soon as I let go, it pinged back by several feet. It was quite clear to me that it wasn't going to reach the mixer. Jake must've known that too, for he had already filled a bucket of water instead.

'Don't worry about that,' he said, on seeing me give up. 'We got it covered, really.'

I gestured at the mixer. 'Perhaps I can handle the cement while you help out your dad.'

Jake reached into the pocket of his jeans. He produced a pack of rolling tobacco and some papers.

'Smoke?'

'I'll pass,' I said. 'It'll only mess with my training programme. You know how it is. Got to give these things a rest every now and then.'

Jake rolled a cigarette for himself with speed and ease. Lighting it up, I watched his chest expand as he pulled on the end.

'So,' he said finally, 'what's with the minipigs?'

The way he said this made it sound like I had opted for the effete version of a real pig. I dismissed his question with my hand.

'It's the wife,' I said, rolling my eyes. 'Anything to keep her happy.'

As soon as Jake's eyes shifted, I knew that Emma was right behind me. I turned to find her looking at me like she had just heard every word.

'Tea's ready,' she said, clutching two mugs in her hand. One belonged to Honey. Father Christmas had given it to her the year before. It was decorated with a wraparound picture of Tinkerbell. The other mug was plain and white. I just knew which one she would offer me. 'So,' she asked Jake, 'are you making good use of my husband?'

'I'm going to be mixing,' I said, before Jake could draw breath. 'Isn't that right?'

Sensing a domestic, Tom's son told Emma that his dad would be needing him.

'One part cement, two parts sand, three parts gravel,' he instructed me, before heading around the house. 'Just be sure to keep it wet, OK?

Emma watched him go, smiling to herself.

'He was talking to me,' I said, before suggesting the kids might be going feral inside.

It wasn't difficult to keep the mix as Jake had asked. Every couple of minutes, I'd slop in a little water and watch the blades churn the porridge around. Even with the sun shining, there was no real heat at that time in the morning. So, it was a surprise to me when Jake returned with the wheelbarrow. I hadn't expected him to be bare-chested, or quite so defined. He really was ridiculously buff, which I guessed must have come from a punishing army life. It didn't unsettle me in any way. I just felt the need to refuse his help when I tipped the first load of cement into the barrow.

'It's no problem,' I said, despite slopping some over the edge. 'The car will grind that into the gravel.'

Jake lifted the barrow, turning it around without losing a

drop. 'We'll need another load in about ten minutes,' he told me.

This time, I really put my back into it. The hardcore wasn't easy to shovel, but I was determined to do a good job. Compared to the kind of work I did at the keyboard, which involved lifting my fingers at most, this was knackering but also quite rewarding. I was very close to breaking a sweat by the time the barrow came back. This time, Tom was pushing it. Not only was his brow slick with perspiration, so too was his bare chest. For a man who was ten years older than me, he was in very good trim indeed. It wasn't the kind of physique you gained from a gym subscription. This kind of toned and sinuous look came from chopping down trees and carrying sleeper timbers on one shoulder. All of a sudden, I felt a bit overdressed.

'With a job like this,' he said, adjusting the shirt he'd tied around his waist, 'the only way to get through it is by not stopping.'

'Absolutely,' I said, unleashing another load of concrete into the barrow as well as across the gravel. 'How many more loads will we need?'

Wiping his brow with his forearm, Tom gestured at the two sacks. We had hardly started on the first.

'All of it,' he said. 'We've marked out a big area.'

'Not the whole garden?' I sounded panic-stricken. 'We're not making a car park here.'

Tom clapped me on the shoulder. 'Relax,' he said. 'You'd be surprised how much slop you need for a job like this.'

'I still feel a bit sad about losing so much of the garden,' I told him. 'How come you don't need hardstanding for your pigs?'

Tom shrugged. 'The bloke who did my inspection didn't

seem that concerned by anything other than the paperwork.'

'Great,' I muttered to myself. 'How lucky I've been.'

'I tell you what, though,' Tom checked we couldn't be overheard. 'If you want me to get rid of the cat bones I see you've stacked in the shed, now would be the time to give me the nod.'

I considered what he meant, and drew breath when I realised what he was getting at. It was a bit shocking as a proposition, but still. So long as the family didn't learn that we had disposed of the cat remains under a bed of concrete, at least there would be no further chance of Roxi and Butch rooting up poor Misty.

'Go for it,' I told him.

'Say no more. We'll keep it between the three of us.'

I waited for Tom to take the barrow around to his son, and then looked around. Emma wasn't watching at the window, and the lane was quiet. Only the dog was nearby, and she was half asleep. It was time to feel like part of the team.

Tentatively, I took off my top. I folded it carefully, and placed it out of range from the mixer.

The air was brisker than I had imagined. It was enough to bring me out in goose bumps as well as the realisation that perhaps my skin tone could benefit from a little sunshine. I stood tall, sucking in the air so my chest drew flush with my stomach, and then breathed out again in one protracted sigh. Feeling suddenly too self-conscious to be comfortable, I decided to pull my shirt back on. Unfortunately, this didn't make me feel any better. All I could do was tell myself to overcome my insecurities and just be relaxed about my body.

A minute later, I removed my top for a second time.

This time I rolled my shoulders, thinking hard labour would help me to warm up, and set about loading up another mix.

Once my body core temperature had lifted by a degree, on account of the physical exertion, it almost felt like a natural way to work. Only when a vehicle pulled up at the gate did I realise this was most definitely a look I'd find hard to justify in public.

With no time to fetch my top, I turned and greeted our postwoman as casually as I could.

'Look at you,' she said cheekily. 'I thought you were just a writer.'

I collected the post from her, thankful that an A4 envelope was among the batch, and clutched it to my chest.

'Just laying down some hardstanding,' I told her. 'Tough job for one man.'

She nodded, as if she understood, and then made way for Jake as he returned with an empty barrow.

'Top moobs,' he said on passing me, before stopping in front of the mixer. 'No time for chatting, though. We need another load back there, pronto.'

With my shirt restored, unlike my self-respect, I worked hard throughout the hour that followed. Each time Tom or his son returned for more cement, I was ready. I learned to spill less on the drive, and even received compliments for the consistency of my mix. Finally, with the last load in the wheelbarrow, I asked if I could help lay the last of the concrete.

'Why not?' said Tom. 'You've earned your stripes today.'

It was my first time away from the mixer since we had begun work. I felt weak from the ankles up as we followed the path around the house. I was too tired to comment on the fact that each journey the wheelbarrow made had chewed a deep and muddy gorge across the lawn. By then, what with the damage caused by the minipigs, I was beyond caring. I just wanted the job done.

As for the hardstanding area, Tom and his son had done a fine job. We found Jake carefully combing the top with what looked like an iron. I'm sure it had a technical name. My only concern was the section near the gate, reserved for the final wheelbarrow load. Butch and Roxi were behind a temporary wall of sheet wire, watching from their ark. Not a single splash of concrete marked their side. My garden had been devastated, first by rooting pigs and now by hard labour. If I was going to live with this, I needed to make my own mark.

'OK,' I said, 'how is this done?'

'With confidence,' said Tom, and invited me to take the wheelbarrow handles. 'It has to go down swiftly and smoothly. 'Otherwise, you'll end up making a big mess.'

Duly, I lined up the barrow. With Tom's nod of approval, I then shoved it out with all my might. I heard him caution me to go a little easier, but by then it was too late. An outpouring of wet concrete overshot the target. Not by much, but enough to push a wave of slop through the area that had been so painstakingly smoothed by Jake. So forceful was the ripple that ran through it several cat bones surfaced in its wake. As I understood it, having watched enough crime documentaries, dark secrets like this weren't supposed to be revealed for decades.

Seeing his hard work ruined, Jake scowled and swore under his breath. My first thought was that he might pull a pistol from the back of his pants and execute me on the spot.

'I can sort that,' I said, uselessly. 'It's no problem.'

Thankfully, Tom was on hand to suggest that Jake might like to wash out the mixer.

'Let's not panic,' he said, as his son skulked away. 'We can get this looking good as new.'

'Do you still need my help?' I asked, as Tom collected the rake he'd been using to push the mix into position.

'You know what?' he said, and looked at me hopefully. 'Jake and I could murder a cup of tea.'

17

The Jail Birds

It took several days for the concrete to harden. Emma took about the same time to soften up having discovered that we had disposed of Misty's remains in it.

She had been the one who ventured out with the cups of tea, just as we were pushing the last visible bones back into the slop. In her eyes, Tom was blameless. For forty-eight hours I was made to feel like I had no respect for the dead.

'You still haven't explained what happened to poor Maggie,' she said at one stage. 'What did you do with the corpse, eh? Dress it in bridal wear and stash it in the loft?'

'That chicken was dear to me,' I countered, unwilling to admit that Tom may well have laid her to rest inside his oven. 'I miss those hens!'

Later, when we let the minipigs back onto the hard-standing, I was pleased and relieved to see no sign of Misty. Together, Emma and I removed the temporary fencing, repositioned their ark upon the concrete, and watched Butch and Roxi inspect their new surface. They showed no awareness whatsoever of the skeletal remains entombed beneath them.

'It looks good,' said Emma, which I took to be an olive branch. 'I know it's been a sacrifice for you, but they do seem to be at home here now.'

'That's because it looks like a small farm,' I replied, smiling all the same.

Emma linked her arm in mine. Butch had found what looked like one of Jake's cigarette papers in the gap between the gate and the concrete. He picked it up between his teeth and scuttled into the ark. Possibly suspecting that her brother was hording food, and essentially driven by jealousy, Roxi followed close behind. Judging by the effort she had to make to squeeze inside, I realised it wouldn't be long before we were calling upon Tom for more help.

'You know what?' said Emma next, and squeezed my arm. 'If it makes you happy why don't you get some more chickens?' She directed my attention to the henhouse and run. 'It's a shame to see it empty, after all.'

At first I took her suggestion to be evidence of clinical insanity. I couldn't even begin to count the number of animals in this household. Then I thought about it some more. In my experience, chickens really were no trouble at all. As foxes had an aversion to pigs, it reminded me of the reason why Butch and Roxi were here in the first place. Above all, I missed a home-made egg on a Sunday morning. There was just one problem with Emma's proposal.

'I don't know anyone who has spare hens,' I told her. 'The last thing I want to do is creep around the village hoping to snatch a runaway from the streets. That's hardly going to help change your view that I'm Dr Crippen to your Dr Doolittle.'

'Why don't you think of buying some legitimately?' she asked. 'You know the big industrial battery farm over the hills from here? Apparently you can purchase old hens really cheaply. They'll still lay, and you'll be giving them a life. That's if you can resist the temptation to murder them this time around.'

* * *

I had only heard talk of the farm's existence. Unlike the free-range set up behind the woods near us, this one existed as a monument to the mass production of chickens. Across the village, people muttered darkly of the industrial scale of the outfit. One of Lou's friends had an older sister who'd found casual work there one summer. She'd lasted a day, quitting because she didn't like to walk between the banks of cages to collect eggs because she ran the risk of being pecked. It sounded like hell on earth for hens.

I knew where to find the place, but it was set so far back from the road that all I could see was a winding concrete farm track. I drove out one afternoon, pulling up at the open gates. There was no sign indicating what lay over the brow. With overgrown hedgerow marking out the way ahead, it looked as if I might be ambushed by backwoodsmen if I ventured any further.

As a signal to any possible patrol that I wasn't here to liberate every bird from the battery farm, I had brought the little ones with me. I wasn't planning on offering them in exchange for my safe passage, using them as a human shield, or anything like that. Frank, Honey and I were just at a loose end after school, and this seemed like a good way to entertain them before teatime.

'Remember, kids,' I said, as we set off along the track, 'this is like chicken prison. It means you have to be on your best behaviour. Otherwise the guards might throw you in the cells.'

Glancing in my rear-view mirror, I saw both Frank and Honey sat on their booster seats looking wide-eyed and worried. I smiled to myself, assured that at least this visit would not be eclipsed by an entirely unnecessary but explosive tantrum.

The track proved to be much longer than I had thought. After driving in second gear for several minutes, it occurred to me that it was way too narrow for me to turn around. If anyone came the other way, there would be a stand-off as to who would reverse first. I pictured encountering some unshaven inbred in a long-billed baseball cap behind the wheel of a pick up, and tightened my grip on the steering wheel. By the time the roof of an industrial shed loomed into view, my expression was close to that of my youngest two children.

'Here we are,' I said, trying to sound as upbeat as I could. As the hedgerow subsided, our view opened up to reveal not one shed but six. Each one had to be the size of a soccer pitch. These windowless giants were arranged in a crescent formation, set within a concrete expanse shot through with weeds and surrounded by rolling woodland and fields. Much to my relief, a hand-painted wooden board ahead read VISITORS. It pointed us towards a car park in front of the nearest shed. Not a single vehicle was in sight. In fact, as I pulled up and switched off the engine, I realised there was no sign of anyone whatsoever. It was also eerily silent. All I could hear was the hum of a generator. Glancing in the mirror, Frank and Honey both had their attention locked on the sheds. All of a sudden, I wondered whether bringing them was such a good idea. Even if we turned around now, I worried that the experience might stay with them for years to come. In a bid to keep things calm, I turned cheerily in my seat to address them. 'Daddy just needs to see if he can find a nice man who can give us some chickens.'

As soon as I cracked open the door, the two children beseeched me not to leave them alone. Panic-stricken, Frank and Honey reached out desperately for me, held back only by their seat belts. Their hysteria only came to an abrupt end when I closed the door in case we were overheard. I turned in my seat. They looked at me fearfully. With my attention fixed on them both, I clicked open the car door once again. With unbridled terror, the little ones shrieked at me to shut it, and fell silent once more when I complied.

'Do you want to come with me?' I asked, unsurprised when Frank and Honey nodded furiously. 'Very well. Just stay close to me, OK?'

All of a sudden, venturing out here didn't feel like a smart move. It was one thing rehousing a runaway from the free-range farm. That seemed like the act of a Good Samaritan. Making our way around this giant shed, seeking out the entrance, I just felt like we were trespassing. What kind of visitors would come here, I asked myself? Several upturned plastic chairs littered a gravel strip between the concrete and the barbed wire. I noted countless cigarette stubs ground into the stones, presumably with the heel of a jackboot.

As a precaution, I checked my mobile phone. My battery had been low all day and I needed to know if it had enough power to make a call. I wasn't sure exactly how that would help us. I suppose I just wanted to feel connected to the outside world. The power bar was fine. The problem, I discovered, was the signal. There wasn't one.

'Isn't this an adventure?' I asked both little ones brightly, to no reply.

Beyond the upturned chairs, a steel door loomed into view. It was about three times the size of a normal door, with a red buzzer fitted beside it. As if designed for visitors who might have trouble recognising what a door looked like, a plate was fixed to the front. It read: DOOR. One could only hope that they could read.

Flanked closely by Honey and Frank, I reached for the buzzer and pressed it once. From inside, what sounded like an air raid siren began to howl. In response, two small hands slipped into mine and clung on tightly. The siren sounded for several seconds, and then fell silent. After half a minute or so, my instinct was to retreat for the car. I could hear nothing on the other side of the door, but for the generator hum. Just as I was about to turn, however, footsteps could be heard approaching from inside. Whoever it was sounded in no hurry whatsoever. Immediately, I pictured some knuckle-dragging giant in a bloodied butcher's apron. On hearing a lock bolt retracting, I added a meat cleaver to the image in my mind.

The door opened outwards, causing all three of us to take a step back.

'Yes?'

The man before us wasn't quite the brawny monster I'd imagined. He was about my height, in fact, ruddy-faced with sandy hair and wearing a collared shirt with his sleeves rolled to the elbow. It wasn't blood-spattered or anything like that. Even so, he didn't give us much of a welcome. Nervously, I cleared my throat.

'Have you got any chickens?' I asked.

He didn't reply straight away. In that moment, as my eyes adjusted to the gloom behind him, and the stench reached my nostrils, the answer became apparent. Under low lights, as far as I could see, stretched rows and rows of stacked cages. Each cage contained a crowd of hens. There must've been thousands, and all of them were squawking wildly.

'Chickens?' The man narrowed his eyes.

Before he marked me down as someone here to insult his intelligence, I explained that we were interested in purchasing a few. Fortunately, this served to loosen the scowl from his brow.

'How many do you want?' He asked. 'Fifty?'

'That might be too many,' I said quickly. 'We were thinking of about, ooh, well . . . about three, actually.'

'Three hens.'

'Four?' I offered, hoping that met his minimum requirement. 'I plan to keep them in our garden,' I added, flustered by the way this was going. 'But there's no problem with the foxes because they'll be guarded by two minipigs.'

'Minipigs.' He looked amused. 'Right.'

Just then, I realised that I might as well have told him the chickens would be protected by ragdolls. With Frank and Honey still clinging to my hands, I had to regain control of the situation somehow. I figured there was only one way forward.

'How much do you want for each chicken?' I asked.

The man considered this for a moment.

'A pound each,' he said. 'Or six quid for the four.'

If his offer was deliberately intended to confuse me, it worked. I drew breath to suggest he may have got his sums wrong, but thought better of it.

'That's a deal,' I said, and wrestled my hand free from Frank in order to find the money.

The man accepted the cash, and told me to wait here for a moment. As he crashed the door shut, sparing us from the smell, I faced both little ones. They were staring at the door without blinking, in a manner that reminded me of our sedated cat.

'Well done for not crying or retching,' I told them. 'I'm really proud of you both.'

A minute later, the door opened once again. This time, the man was accompanied by a colleague. This one was a closer match to the knuckle-dragger of my imagination. He wasn't bloodstained, but he certainly looked capable of wholesale slaughter.

'Here are your hens,' the man said, and stood aside so I could collect them.

At first I thought he was making a sick joke, because the big guy before me was clutching all four chickens by their legs. Dangling and flapping uselessly, the poor things were kicking up a dreadful noise. Then he gestured for me to take them from him, which caused me to retreat further still.

'Don't you have a cardboard box or something?' I asked.

The man with the rolled-up shirtsleeves looked like he was running out of patience with me.

'That'll be another quid,' he said, and turned to fetch one.

Unlike the minipigs, I did at least have some experience of caring for chickens.

Once we were home, I placed the box containing the four new arrivals in the shade beside the woodpile. In such close confines, so long as everything was quiet, hens underwent a weird kind of shut down. In this calm and dozy state it was quite alright to leave them until dusk, which is always the best time to introduce them to a new home. It meant they roosted on instinct, the chicken equivalent of sleep, and hopefully bonded in the process. So it was that under an early starlight, I crept down to the pigsty with the box, opened up the porthole on the side of the henhouse and popped them in one by one. It was too dark to really get a good look at them. I was just grateful that they didn't flap, and simply settled in for the night. As I replaced the porthole, I spotted Roxi's snout emerge from the straw inside the ark. She sniffed a few times, before retreating back to her bed. I took this to be an encouraging sign. I hoped it meant that the minipigs would be entirely accommodating when it came to sharing their space.

Daybreak, as ever, was marked by the sound of birdsong.

One thing was missing, however. After weeks of waking into a frantic scramble for the pigsty, I wasn't summoned by the sound of minipigs squealing. Lying in bed, I heard not a single grunt that could potentially contravene noise abatement laws.

'Something must be wrong,' I said, up on my elbows now.

Emma joined me, yawning into the back of her hand, and then tipped her head to listen.

'Maybe they're having a lie in,' she said. 'Minipigs must go through an adolescent phase like dogs and cats. Just so long as they don't start slamming doors and sulking, that's fine by me.'

Emma flopped back onto her pillow. I wasn't so assured.

'It's not right,' I muttered, and padded to the window. Moving the curtains aside, I took one look and suggested Emma might like to do the same.

'What is it?' she grumbled before drawing beside me. 'OK, that's weird.'

Outside, in the pigsty, Butch and Roxi were certainly wide awake. But instead of bellowing for their breakfast, they were stationed on each side of the henhouse. Both stood quite still, their attention locked onto the air vents in the chickens' new home. Faintly, I could hear the sound of clucking from inside. As far as I was concerned, the silence of the minipigs was an unexpected gift. It meant I didn't have to rush outside with my dressing gown flailing to shut them up with breakfast.

'It is weird,' I agreed, and sauntered back to bed for the first lie in since the minipigs had arrived, 'but also slightly wonderful.'

Normally, I would be met at the gate by two very hungry animals. When I finally went out to feed them, I entered without being obstructed. Butch and Roxi didn't even acknowledge my presence. Both were absolutely rooted to the spot, evidently straining to pick up on every cluck and warble from within and work out what was behind it. They looked like they'd been put on hold in some way, which was probably wishful thinking on my behalf. Had the minipigs come fitted with a pause button, they'd probably still have been living in the house.

'Are you ready to meet the new housemates?' I asked, stepping around Roxi so that I could open up the henhouse into the coop. 'Be nice, OK? They've not had the greatest start in life.'

With a twist of the door handle, I opened up the henhouse. My last lot of chickens would always bundle out flexing their wings after a night inside. This time, there was no sign of the new occupants. Crouching behind Roxi, who was still staring at the vents, I peered inside.

Then, one by one and looking utterly awed, four wraith-like hens' heads extended outwards. I had heard that battery chickens tended to have long necks, on account of having to stretch down to the feed belt. I just wasn't prepared for the stalks on these birds to practically intertwine like snakes.

Butch and Roxi looked as stunned as I felt. They only moved when the first chicken stumbled from the roost, followed by the others. Startled and unnerved, the minipigs scampered behind me as if hoping for protection. If they hadn't reacted first, I would've been the one who retreated behind the minipigs, for the wretched creatures struggling to their feet before us were a truly unsettling sight. Having only glimpsed the hens in the gloom at the battery farm, and then handled them at night, I hadn't realised just what a state they were in. All four were scrawny in build and missing so many feathers that they looked like they were half ready for the roasting dish. Beneath the quills, their skin was a deathly pale, as I suppose it would be given that they hadn't once seen sunshine in their lives. So shocked was I by their condition that I didn't hear Lou and May enter the pigsty until they were right behind me.

'Eew!' This was Lou, the first to recoil in horror. 'Dad, they're gross!'

'Are you sure they're chickens?' asked May, and crouched beside me for a better look. 'They don't look very cat friendly.'

As we looked on, the four hens became aware that they were under scrutiny from both man and minipig. Evidently, they didn't like it one bit.

'What are they doing?' asked Lou, as several chickens turned their attention towards us and squawked furiously. Not only were they short of a few feathers, I wondered whether they might have been missing a few brain cells, too. They looked like they were ready to pick a fight with us just for staring at them strangely. At the far end of their run were two hoppers containing food and water. Unlike any chicken I had kept in the past, they were showing no interest in them whatsoever.

'These hens have never experienced freedom,' I said. 'I can only think it's going to be a while before they realise that they don't need to act like tough guys in order to survive.'

Lou turned to me. 'Are you suggesting we need to show them love?'

'They won't get that from Miso,' her sister added, rising to her feet. 'Maybe the minipigs can help them socialise.'

By now, all four birds were eyeballing us, flapping their wings and making such a noise I worried that people might think we'd got into cockfighting. Behind me, Butch and Roxi looked both silenced and terrified.

'I think for now we'll keep the chickens in their run,' I said. 'They just need a little time to settle.'

* * *

Inside, I found the little ones eating cereal from bowls brimming with milk. Battle ready for work, Emma was scrolling through emails on her BlackBerry. It seemed to go on endlessly. Sometimes I wondered what it would be like to get more than two emails a day. From people other than my spouse.

'Looks like the chickens might be a project,' I said, slipping off my boots for the dog to inspect.

Lou and May followed me inside. As May headed off to find the cat and check him for vital signs, her sister stood beside Sesi looking utterly indignant.

'A project?' she looked stricken. 'Mum, it's a social *disaster*! I can't bring my friends here to see the minipigs any more. Not with those minging hens in the pigsty.'

'Of course not.' Her mother made a vague effort to look like she was listening, despite the fact that her thumb was still turning the scroll wheel and her eyes were locked on the screen.

'*Mum*,' said Lou, more insistently this time. 'You need to talk to Dad. He's under the impression he can save them. It's like he thinks he's the chicken messiah or something.'

'That's nice'

'Emma!' This time I drew her attention. 'Did you hear what she just said?'

My wife looked at me blankly, which was an answer in itself.

'Sorry,' she said to us both now. 'I've got a lot of stuff to deal with at work today. Can't you sort this out between yourselves?'

I drew breath to state that I had no problem with bringing our eldest down a peg or two. Before I could utter a word, Emma tutted at her screen.

'Problem?' I asked, resigned to the fact that her BlackBerry was about the most important thing in the world to her inside office hours.

'It's a reply from the vet,' she muttered. 'I was hoping they could visit on a weekend so that I could be here.'

'For what?' I asked, thinking at the same time that at least she was speaking now.

Finally, Emma looked up at me.

'They're coming to castrate Butch,' she said. 'It's just there's no way I can take time off work at the moment.' With a sigh, Emma tossed her BlackBerry into her bag, collected her coat from the back of the chair and then kissed the little ones goodbye. 'Will you be here on Friday?' she asked. 'That's the earliest they can do it.'

'Let me think about that.' It was only right that I gave this some consideration, simply so the children might believe their father actually had more going on in his life than just pet-sitting and preparing the tea. 'I should be here, yes.'

Emma beamed at me from the back door.

'Thank you,' she said. 'You know, I'm genuinely sad to be missing out. It sounds fascinating. Apparently when the operation is performed on site they always need a helping hand.'

18

A Bump in the Night

Butch's fate wasn't something I liked to think about. Having faced a less drastic procedure myself, to achieve roughly the same objective, my memory of the experience still made me wince. Besides, my mind was preoccupied by other concerns.

The first was the number of bin bags stacking up outside the gate. Cleaning out the minipig muck was an effective way to keep the fly count down. It's just my bid to give the stuff away wasn't going well. In fact, it wasn't going anywhere. I had thought there would be a strong market in the village for a substance guaranteed to give a kick to any manure heap. Instead, with no takers, we had just earned ourselves a reputation as the 'poo house'. As soon as Honey reported what her friends at school had said, I received clear instruction to remove the sign and the bags from public view and find some other way of disposing it.

Tom made just one suggestion. When I asked for his advice, his response wasn't terribly helpful.

'Give it back to the minipigs,' he said. 'It's their mess. Let them deal with it.'

'I'm serious,' I said. 'Butch and Roxi don't care. They just tread in the stuff.'

'And that's exactly what they should be doing. It's nature at work.' We were standing in the yard at Tom's smallholding.

I had been out walking with Sesi when I decided that I needed his advice. The dog was sitting obediently, even if it was beside Tom and not me. All around us, animals existed in harmony with their surroundings. The horses grazed out in the paddock, hens pecked happily at our feet, while out in the pigs' field the new arrivals were basking in the sunshine. I had found Tom filling a bucket from a water butt. Quite literally, he had everything here on tap. 'Do you see farmers picking up cow pats?' he asked me. 'No, you don't. That's because there's no need, and the same goes for pig dung. It just gets trodden into the ground and decomposes naturally . . . unlike your cat,' he added after a moment.

Despite my best efforts, I couldn't help but smile. 'But if I leave it on the ground we have a fly problem.'

'You're a farmer now,' he reminded me. 'Flies are a hazard of the profession.'

I looked around, saw no hint of the dark, pestilent cloud I could expect if I followed Tom's advice.

'Why don't you have a problem? I asked.

Tom shrugged. 'There's a lot more room for them to buzz around here. If it's that bad up at your place I should imagine Emma will be making the case to bring the minipigs back inside.'

'Don't go there,' I told him. 'At the moment, it isn't an issue for anyone else but me. The flies are only drawn out by the heat of the day, and Emma is at work then. They've bothered her at weekends, but by Monday morning she has other things on her mind. Until they start crawling across her BlackBerry screen, it's my problem and not hers.'

Tom collected the bucket full of water.

'Flies are seasonal. They won't be around all year,' he said. 'Unlike the rats.'

I looked at my feet, and then breathed out long and hard. Everyone else recognised the fun to be had from keeping minipigs. I just saw the trials.

'What can I do about rats?' I asked, sounding as resigned as I felt.

'I wouldn't worry,' he said. 'You have one cat left, don't you?'

'In the loosest sense,' I replied. 'If any kind of vermin scuttled across Miso's field of vision, the instinct to pounce would take several hours to reach his brain.'

'OK, let's hope the rats steer clear. Somehow, you need to find a way to enjoy your pigs.' Tom gestured at his livestock. 'I don't keep mine as pets, but raising them is something I really enjoy. Just look at them. They're gentle giants. As I see it, pig keeping is good for the soul.'

'But not for the garden,' I added. Even so, I thought about what he had said. 'I'd really like to think one day I could take a break from my office, amble down to see Butch and Roxi and come back feeling chilled. Unfortunately, there's no chance of that at the moment, which is rather my fault.'

Tom looked puzzled. I told him about the new chickens.

'Escapees from the free-range farm? They're no trouble at all.'

'I got this bunch from somewhere else.'

Tom's expression darkened considerably. 'You didn't go to the battery sheds, did you?'

I nodded like a small boy in the wrong. 'I had no idea they'd be so vicious,' I told him. 'I thought I was doing a good thing.'

'You are,' he cut in. 'Giving those birds a chance to live freely is a fine and noble act. The trouble is they've spent their whole lives behind bars. It's a tough life in there, and suddenly you've sprung them free. If you'd done the same thing to a bunch of lifers, they wouldn't come out as models of society, would they? They've been institutionalised, and old habits die hard.'

I felt crestfallen. Everything Tom had said made sense. 'They certainly look mean,' I said. 'It doesn't help that half their feathers are missing, but there's a cold, rock hard look in their eyes that scares me.'

'It's a question of rehabilitation,' Tom assured me. 'What do the minipigs make of them?'

'At first they were curious. Then they were terrified. Now it's just got silly, and it's all because of the eggs.'

'How so?' asked Tom, intrigued.

'Whenever a hen lays one in the nest and then makes a song and dance about it, Butch and Roxi go crazy. It's like watching crack addicts trying to break into a doctor's surgery. The chickens don't like it one bit. The trouble is I don't want to keep them in the coop. The reason I got them was to give them some freedom. I'd really like to see them roaming the pigsty, but not at the cost of my eggs.'

To my surprise, Tom brightened at this.

'Give me ten minutes,' he said, and invited me to follow him around the stable to his workshop. 'I think I can build you the solution.'

Later that day, trying my best to ignore the flies, I stood in front of my hens' coop and prepared to open it.

'Check it out,' I said to the minipigs at my feet. 'I'm surprised I didn't think of this myself.'

From behind the fence to the pigsty came a long sigh. 'Even if you'd thought of it, Dad, you'd never have been able to make it.'

I turned to face Lou. She had come out with me, keen to see if the device would work.

'I may not share Tom's skills with woodwork,' I told her, 'but even *I* could've made one of these.'

'Sure you could,' replied Lou. 'If we actually owned a saw and a hammer.'

She had a point, and in any case I was unwilling to mark this moment with an argument, so I crouched to check that everything was secure. What Tom had created was incredibly simple but guaranteed to allow my chickens to roam freely without fear of losing their eggs to the minipigs. He'd cut a length of fencing trellis in two, and then nailed them together in a 'V' shape. All I'd had to do was place it on its side in front of the coop, and secure it to the mesh with the strips of wire he'd cut for me. This created a kind of Green Zone for the hens. With the door open, they could clear the fence with a flap of their wings, unlike our diminutive pigs. Roxi may have been able to peer longingly over the top, but even I knew the minipigs couldn't fly.

'There,' I said, and pinned open the coop. 'Freedom at last.'

The chickens wasted no time in seizing the opportunity. First one hopped over, followed by another, in the manner of a jailbreak. Instead of making a run for it, however, the birds seemed to come out spoiling for a fight. Within seconds, Butch, Roxi and I found ourselves retreating from my four menacing hens.

'Is this a good idea?' asked Lou. 'They look like they're concealing homemade weapons.'

Judging by the way they fanned out and prowled around us, it really did look like one of them might whip out a razor blade strapped to the top of a toothbrush.

'Tom told me not to worry. He said they'd sort it out among themselves.'

I glanced down at Butch and Roxi. Having initially regrouped in front of their ark, both minipigs were showing more courage than me. With twitching snouts, they began to inch towards the chickens.

'Do they have names?' Lou asked.

'Not yet,' I replied, watching one bird eyeball Roxi. 'Tom doesn't personalise his pigs to make it easier at slaughter time. I think it's best not to bond with them until we're sure they're here to stay.'

'You can't do that,' she replied, sounding outraged. 'What's the point of rescuing battery hens if you're going to *murder* them?'

I looked across at my daughter. 'That's a bit extreme, Lou. I just mean that we might need to find another home for them if it fails to work out with the minipigs.'

Lou tightened her gaze on me. 'A pet is for life,' she reminded me. 'Even the manky ones.'

Sensing further grief if this conversation continued, I returned my attention to the hens. They were still staking us out, but rather than direct hostilities they appeared content with just sizing us up. Even when the minipigs ventured closer, they ruffled what plumage still clung to their backs but held off from an attack. With great relief, a first for me, I watched Roxi plough her snout into the ground and devote her attention to digging. Butch focused on the anti-minipig device in front of the coop. He gave it a little shove with his snout. When it didn't move, he turned and joined his sister. At the same time, the chickens stood down and began scratching at the soil.

Lou looked at me. I beamed at her, nodding at the same time.

'Tom was right,' I said. 'This might just be a happy place to hang out.'

My eldest daughter pulled a face. 'Not until those birds grow some feathers,' she said. 'I've got a list of people from school who want to pet the minipigs, Dad. But that's not going to happen until the chickens look like, well, *chickens*.'

I glanced down at Butch as she said this, mindful of the procedure he faced before the week was over.

'This weekend,' I told her, 'you can bring them both inside. Butch will probably welcome a sofa to sit on.'

'Great! Can we make it a sleepover?'

'I suppose it must be our turn,' I said with a sigh. 'But the minipigs can't stay the night, OK?'

This was to be the third occasion our house had been host overnight to a flock of teenage girls. I've no idea what the correct term is for a bunch of boy-crazed 14-year-old females. All I know is that their group behaviour involved flitting en masse throughout the house, twittering as they went.

When I informed Emma that our weekend was set to be ruined, she seemed as resigned as I was to the fact that our time had come.

'It'll be fine so long as we get some sleep over the next few nights,' she said. 'Who knows? Maybe your minipigs will turn a corner and give us an extra half hour in the mornings?'

We were lying in bed at the time. Emma was talking to me while reading her book.

'*My* minipigs?' I looked across at her. 'Since when were they my responsibility?'

'Sorry, did I say they were yours?' Emma turned a page, read for another second before dropping the book to the floor. 'My mistake,' she added, before reaching up to kill the lights.

It took me much longer than usual to nod off that night. My wife once claimed she could sleep within moments because she'd been working hard all day. Even she realised what thin ice she had strayed upon with that statement, and never went there again. Her comment about the ownership of Butch and Roxi might've been a slip of the tongue. Even so, it left me staring at the ceiling for some time.

I can't say when sleep finally beckoned. All I know is that I was awoken shortly afterwards by a thud. It was loud enough to make the dog start barking downstairs.

'What was that?' I hissed.

Emma murmured something about a unicorn in the elevator, which strongly suggested she was still in the grip of sleep. I lay there for a moment, straining to listen. I kept a weapon under the bed as a security measure. I'd have felt better if it was a handgun, or even a baseball bat. Strictly speaking, Frank's old plastic light sabre wasn't going to present much of a threat. For one thing it was out of batteries. I just hoped that if I ever had to wield it in the dark it might pass as something Samurai.

Finally, Sesi fell quiet. A minute later, the same muffled crashing sound caused her to kick off again. This time it was clearly coming from outside. Creeping from the bed, I crossed the floor and peered through the curtains. A full moon silvered the garden. I saw no sign of anything untoward, which was a relief. The next time I heard it, I knew for sure that it was minipig-related. I just couldn't work out what was going on inside their ark. All I could do was return to bed, and try to get the sleep I so desperately needed if I was going to survive the weekend.

Throughout that night I was snatched awake by the same abrupt disturbance. Each time, Sesi would react and yet Emma did not stir. Towards dawn, ragged from exhaustion, I decided it was only fair to rouse my wife so she could hear it for herself.

'Just listen,' I said. 'It's happening every minute or so at the moment.'

Sitting up with her eyes crinkled shut and her hair astray, Emma angled her head to one side and waited. The thud, when it came, was loud enough to vibrate the windowpane. Emma sighed to herself, and flopped back on the pillow.

'It's just Roxi turning in her ark,' she said, and hauled the duvet over her. 'You did ask Tom about building a bigger one, didn't you?'

Now she had put the sound into words, it was unmistakeable. For the next half an hour, listening to the minipig struggling to get comfortable, I tried to calculate how much notice my friend would need before he could create something more accommodating. Finally, Roxi fell quiet long enough for me to settle. By now, dawn was on the cusp of breaking. The first notes of birdsong should've been enough to lull me into a late sleep. As it was, I lay there in a state of rising anxiety, braced for the first hungry squeals from the foot of our garden to drag me into the day.

That morning at my desk, I felt washed out and unfocused. The weekend was almost upon us. I should've used this as an incentive to knuckle down. Instead, knowing I would get no rest, I just pecked out words on the screen. On the upside, it seemed like the Green Zone was holding up. Having rushed down to feed Butch and Roxi and open up the henhouse, both Team Minipigs and Team Chicken appeared untroubled by each other. Food was what kept them in their respective corners, of course. Even so, it genuinely seemed to me that the tension between them was over.

With my window open I could even hear the hens clucking. It felt good to think that had I not braved visiting the battery farm with the little ones, those four birds would still be imprisoned in miserable cages. Outside, they could look forward to an existence that offered them space, natural light and peace, and hopefully no fear of a visit from any predators. From my previous experience of poultry keeping, I even learned to recognise when a hen had laid an egg. Their chirping became quite song-like, which was always nice to hear. So, it alarmed me when the sound of one such happy hen suddenly turned nasty. The clucks became squawks and frenzied cries. Within seconds, I was out of my office and barrelling for the back garden with Sesi leading the way.

'What's going on?' I called out, for if a fox had pounced my voice might send it fleeing. 'I'm on my way!'

It was a relief to see that nothing feral was behind the commotion. Then again, seeing Roxi inside the coop was both disappointing and irritating in equal measure. Somehow, she had struggled over the top of the trellis barrier and headed straight for the henhouse. I knew this for sure because her mouth was crusted in eggshell as she chomped away. Understandably, all four hens had rounded on her. They flapped and pecked at her, making an almighty noise, but the minipig seemed oblivious. I can only think the sheer joy of tasting yolk for the first time made it a price worth paying. What was unnecessary, I thought, was the fact that Roxi made a temporary stop in her bid to escape in order to deposit a poop in the coop.

'Oh, come on,' I moaned, letting myself into the pigsty. 'What are you? A teenage burglar? Get out of there!'

Watching Roxi make her undignified exit, chased out by the chickens, I realised I would have to raise the height of the Green Zone. Butch had witnessed the entire incident from the outside, unable to clear the trellis as it currently stood. He looked most disgruntled as his sister fled from the attention of the ex-batts. Judging by the way they reluctantly gave up on their quarry, shooting stares and clucks in Roxi's direction, they looked like chickens who could hold a grudge.

'Right,' I muttered to myself. 'It's time to make sure this doesn't happen again.'

A search inside the shed revealed that I had no spare timber or chicken wire to raise the height of the Green Zone. Unlike Tom's workshop, this space was just crammed with unwanted junk. Still, I was determined to be creative here, and use what I could find. Tom would be proud of me, I thought, as I searched for suitable materials.

Later that day, after school had finished, the fruits of my labour were the first thing the children noticed. Normally they would pour from the car and head straight down to say hello to Butch and Roxi. This time, they stopped at the garden gate and just stared.

'What's happened?' asked May, and faced me looking a little distressed. 'What have you done?'

'Used my initiative,' I said, standing behind the little ones to admire my work. 'I think it's clever.'

'Dad,' said Lou in a way that made me think she was about to kick off. 'You need to take it all down before Mum gets back.'

'Why?'

'Because it looks *hideous*!' she snapped back. 'It's garish enough to be seen from space, for crying out loud. What if the Google Earth satellite took a snap now? People would think we have new age travellers camping out in our garden.'

'Now you're overreacting.' I opened the gate and invited them all to take a closer look. 'I've simply recycled some things in order to keep the minipigs out of the chicken coop.'

As Honey and Frank scampered down the garden, their older sisters just stood their ground.

'Are those our body boards?' asked May. 'You should've asked me first. They're for summer holidays, not hens.'

'Same goes for the swing ball set,' Lou added. 'It isn't designed to be part of a barricade.'

By now, the little ones were crouching at the fence, petting Butch and Roxi. Two of the chickens looked down on them. They were perched on my creation, showing the same quiet menace as nightclub bouncers. What I had done was raise the barricade that Tom had built from two lengths of trellis. With no nails or twine to hand, I had managed to keep the body boards in place between a series of posts. Ideally, I'd have used canes. As I possessed none, the pole from a swingball set had proved a good substitute, along with three shrimping nets, a garden hoe, the oars from a dinghy that had punctured long ago, as well as two blue plastic tennis rackets jammed against the boards at an angle. As a precaution, I had underpinned the whole construction with a series of bright-coloured tent pegs, and then bound the entire thing together using the string from a kite, still attached, which I had tied to the top of the coop as a kind of anchor.

The minipigs had watched me at work with some interest. When I had finished I waited for them to sniff around the boundary, before leaving in the knowledge that the hens could lay their eggs in peace. I just hadn't anticipated a level of objection from my children normally reserved for plans to build one thousand homes on ancient woodland. By the time Emma returned from work, she found Lou and May all but chained to the front gate in protest

'Please, Mum! Just tell him!' This was Lou, whose primary concern was what her friends would say at the sleepover. 'You can ask Tom to do it properly.'

'So it looks nice,' added May, clutching Miso for comfort. 'And not like a car boot sale.'

Emma had yet to take her coat off. It looked like she had suffered a long day at the office.

'Do I need to see this?' she asked me.

I told her there was no immediate rush.

'It serves a purpose,' I added. 'We've had two eggs since I built it.'

Briefly, she looked impressed. Unfortunately, Lou and May persisted.

'Please, Mum. It's a colossal bodge job. Next thing, he'll start cementing it with mashed potato or something.'

As they all left me to look out of the front room window, I sighed to myself and awaited Emma's response. On hearing nothing, I followed them in. Emma was standing at the glass, flanked by Lou and May. Nobody said a word.

'What do you fancy for supper tonight?' I asked, anxious to break the silence. 'Omelette?'

Grim-faced, Emma turned to face me. 'How long did it take you to build?' she asked.

'Not long,' I said with a shrug. 'Unless you're counting adjustments, in which case most of the day. But the fact is it works.'

I was braced for her to advise me that I needed to call Tom. Instead, she placed her hands upon the shoulders of the girls at her side and told them to prepare a salad to go with supper.

'At least you make a good omelette,' she said, and offered me a smile.

May and Lou looked appalled. '*Mum!*'

'Your father's gone to time and trouble to keep the peace in the pigsty. It's not for us to criticise.'

'Really?' I couldn't quite believe that Emma had accepted my handiwork. Lou seemed even more flabbergasted.

'You're joking, right? What am I going to say to Jade, Lauren,

Kath, Darcy and Sara? They'll take one look and think my dad has turned into the sort of confused old man who loads the dishwasher with dirty clothes. It's embarrassing, Mum. I'd rather die!'

'Instead of death,' replied Emma, who seemed to be tiring of Lou's attitude, 'you can just postpone the sleepover by a week.'

Lou's lower jaw fell as if she'd lost all control of it. 'I can't believe I'm hearing this!' she shrilled at her mother, before jabbing a finger at me. 'You always take pity on *him*!'

Judging by the way Emma shooed Lou and her sister into the kitchen, it genuinely seemed that I had her support.

'That's great,' I said. 'It's always nice when good work is appreciated.'

'I wouldn't go that far,' she said quickly. 'I'm just aware that you didn't get much sleep last night thanks to Roxi. The fact is she's going to keep tossing and turning until Tom has time to build them a new ark.'

'I could always do that,' I suggested, only for my confidence to wither when Emma looked at me in amusement.

'Listen,' she continued, and took my hands, 'you've made a big effort with these minipigs, and I'm grateful. It means a great deal to me. Really it does.' She kissed me on the lips, which came as a surprise, then pulled away to look at me. 'Besides, I'm not the one who has to explain what you've built when the vet comes tomorrow.'

19

Cut to the Chase

It was a smart move. Had Emma kicked off about my DIY skills, I would've stood my ground much like the ex-battery hens. I accepted that the reinforced Green Zone didn't look like a work of art, unless of course we were talking abstract modern art. Nevertheless, I had built it with my own hands. At first, this is what appeared to count for her. Then Emma reminded me about Butch's appointment with fate. From that moment on, I considered the construction from a vet's point of view.

That night, instead of taking some pride in the fact that I had actually made something which hadn't failed or fallen apart, I cringed at the prospect of presenting it to a professional in the field. Emma had contacted a veterinary surgery who specialised in dealing with livestock, rather than one closer to home with more experience in running over cats. I imagined that whoever they dispatched spent their days visiting proper farms. A pig raised at a domestic address, in a space containing the contents of the shed, basically marked me out as eccentric at best. Even so, despite the potential for embarrassment, I remained determined not to cave in and take it down. The system worked in the best interests of both minipig and chicken, and any vet would recognise this.

In my office the next morning, as I squeezed in some

work before the appointed hour, I heard nothing outside but contented clucking. Such pastoral strains should've helped to relax me. In reality, since I had been volunteered by my wife to assist in Butch's castration, I worked with a tension headache up to the moment that the doorbell rang. When I shuffled out to answer it, I should say that the pain between my temples wasn't the reason I had to blink for focus.

'Mr Whyman? I'm here to give your pig the snip.'

I don't know what I was expecting the vet to look like. One thing is for sure, I could never have imagined she would shape up like this. Fundamentally, everything about her was in direct contrast to the lady from Trading Standards. The woman before me, speaking with a German or Austrian accent, had mile-high cheekbones and honey-coloured hair that tumbled over her shoulders. She was wearing heels, a pencil skirt and blouse that really should've been buttoned a bit higher. A little part of me wondered whether I was missing something here. Even if this was my birthday, Emma wasn't the sort of wife who would hire me a stripper. With a gleam in her eye, my visitor offered me her hand to shake. When I reached for it she pulled back and made a scissoring gesture with two fingers. 'I hope you're cut out to help me?'

'Sure,' I said hesitantly, and cleared my throat. 'If you think you need my assistance.'

This time, with a half-skewed smile, she shook my hand in greeting. 'It'll be emotional,' she said. 'Especially for the pig.'

'It's a *mini*pig,' I told her, when she finally released my hand, but I don't think she heard me.

'If you'll just give me a minute to get my things, we can begin.'

'Great,' I said. 'I'll just grab my boots.'

As I hurried back through the house, I wondered whether

this visiting vet was properly qualified. Above and beyond her outfit, the whole thing with the handshake had really taken me by surprise. If my doctor had pulled such a stunt when I showed up for the snip, I'd have turned full circle and never returned. As it was, she was here at our house, getting ready to enrol Butch in a club from which we could never return.

'Sesi, you had better stay here,' I said, on reaching over the child gate to collect my boots. 'This is a delicate procedure. No dogs allowed.'

Returning to the front door, I decided to head out and find the vet. I figured there must be quite a lot of equipment to carry, and wanted to lend my support. From the path I could see her car reversed into the drive. The boot was open, as was a steel case inside. Curious, I approached to take a better look. The case contained an array of surgical tools and medicines.

'That's some equipment you have here,' I said, aware that she was busy doing something behind the driver's side of the vehicle. I remember glancing over the back seat as I spoke, and finding her midriff filling the side window, bare but for the cups of a lingerie bra.

'I'll be right over,' she called back, upon which I realised she was hauling herself into what appeared to be an all-in-one scrub suit. When she did emerge from behind the car, tying her hair back with a rubber band, she found me staring at the gravel, all but incinerating from the neck up. 'Everything OK?'

'Fine,' I said, and switched my focus to the surgical box. 'Would you like me to take this?'

Effortlessly, she swung it from the car. 'I can manage,' she said, and invited me to lead the way. 'But I'll need your assistance a little later.'

At the best of times, I struggle with small talk. Having spent my career working alone or communicating with children and animals, I have always lacked the ability to conjure a conversation out of nowhere. On this occasion, I wasn't just lost for words because the vet who followed me down the garden path was smoking hot and happy to undress in my drive. What really caught my tongue was the explosion of squawks from the pigsty. I had been prepared to explain away my set-up to protect the chicken coop. On reaching the garden gate, however, it became quite clear that it was now unfit for purpose.

'What is that?' she asked, drawing alongside me.

'A minipig,' I replied, as Roxi bundled her way back over what remained of the barricade, munching on an egg and pursued by angry hens. 'The other one is really cute.'

'I'm not talking about the pig,' she said, upon which I realised what she meant.

Following the vet towards the pigsty, batting away the first of the flies, I wished Roxi had chosen to destroy the Green Zone long before this moment. At least then I could've tidied away all evidence that I was hopeless with my hands. Instead, as Roxi hurtled from the hens with the kite string wrapped around one hind leg, the parade of tangled-up junk that followed her told its own story. With everything stripped away, from the body board to the plastic rackets, only Tom's original structure held true.

'I thought I was being resourceful,' I told her.

'I'd like to say I've seen worse,' she said. 'Now, where is the patient?'

Throughout, Butch had been watching from the ark. Being closer to the ground than his sister, he was wary of the hens. Especially when they were angry.

'This is the patient,' I said. 'I haven't told him what's in store.'

The vet let herself into the pigsty. Having swapped her heels for trainers, she set her case up on the hardstanding and then deftly stepped on the kite string as Roxi passed. It was enough to stop the minipig in her tracks. Just long enough for the vet to uncobble her from the string.

'You can clear this up now,' she told me, in a way that felt like an order. 'It'll give me a moment to prepare.'

I could only hope things would improve once the procedure was underway. I wasn't exactly looking forward to it, especially as I had no idea what my role would be. As the person charged with preparing the operating theatre, it took me five minutes to stack the junk, ready to return it to the shed. I also decided the hens should be confined to the coop. Just so we could work without fear of being mugged. As I closed them in, the vet confirmed she was ready by squeezing a little liquid from the tip of a big syringe she had prepared.

'I'm sure you appreciate why we need to use a sedative,' she said. 'But first you must catch him.'

'Me?'

'For sure.'

I glanced at Butch. He was watching us intently, but hadn't moved from the ark. His sister, meanwhile, was busy rearranging one of the many craters she had made in what was once my garden.

'I'm game,' I said, and cracked the knuckles in each hand. 'Let's do this.'

I made my approach, inching forward with my hands outstretched. At once Butch's ears pricked and he retreated into the ark.

The vet tutted at me. 'You can't creep up on a pig. They're not stupid. You have to *dive* on them.'

I looked at the ark. There was no way I could even fit my

head through the opening, let alone charge in and make a grab for him. With no other choice, I made my way to the back of the ark and lifted it by several inches. Sure enough, with a protracted squeal Butch slid out and scrambled for freedom. He dashed around the chicken coop, only for his passage to be blocked by the vet. On doubling back, he found me throwing myself at him. I suppose it wasn't hard for him to sidestep out of the way. In vain I tried to grab his tail, only to catch my foot in one of Roxi's excavations.

'This could be a challenge,' I said, on picking myself up. I turned to look for my quarry. Butch was some distance away, standing side on with his eyes locked upon me. I felt like I had just betrayed him in a way that he would never forgive. I spread my hands, thinking perhaps I could reason with him. 'Look, it isn't going to hurt. Not once you've had the jab. I promise you. I've had this procedure myself.'

'A castration?' The vet sounded surprised. 'A chemical one, surely?'

I turned on the spot, aware that she was probably marking me down as some kind of recovering sex offender.

'I'm not castrated,' I said, struggling to find the right words. 'I meant I've had a vasectomy. Between males, it's kind of the same deal.'

She seemed to accept my explanation, still clutching the syringe, and then switched her focus to a point behind me. 'Be ready,' she said under her breath. 'He's approaching you.'

I glanced around. Sure enough, Butch was climbing over the craters towards me. It was as if he could not believe that I would ever do anything bad to him, which made me feel so much worse as I braced myself to fall upon him.

A moment later, I found myself horizontal in the dirt again. The only difference was that now I had a minipig underneath me. One making such a noise I could barely hear my own voice.

'Quick!' I yelled, wrestling to keep Butch in my clutches. '*Let him have it*!'

I probably didn't need to tell the vet what to do. She was at my side in a flash, and though I didn't see her sink the needle into Butch, the spike in his screams told me we had been successful. As soon as I let him go he made a beeline for his sister, and jammed himself between her flank and the chicken coop. Rising to my feet, with the vet at my side, I drew breath to offer my congratulations. Instead, I found my attention drawn to a creak of the fence behind me. What I saw first were a pair of hands clad in gardening gloves, and then a fearful-looking face peeping over the top.

'Roddie!' I said, in a bid to sound as pleasantly surprised as I could on facing our neighbour at this moment. 'What brings you down here?'

As ever, and despite his gardening duties, Roddie was dressed in his golf club blazer and tie. With port-flushed cheeks and a head of white hair, he looked like a schoolboy who had aged prematurely. 'I have been collecting the first fall of apples from the orchard.' My neighbour shifted his gaze to the woman in the scrub suit beside me. 'What is going on?'

I realised that it looked more like we had just enjoyed a boisterous round of dirt wrestling than completed the preliminary stage of a surgical procedure. It also didn't help that Butch was beneath Roddie's line of vision. The little minipig was still close to his sister but seemed increasingly weak on his feet. Roddie, meanwhile, had clearly made his presence known on account of all the noise.

'The racket I just heard nearly gave me a heart attack. I am growing tired of such awful squeals. Whatever you're doing,

I would be grateful if you conducted yourselves with some respect for those within earshot.'

'Of course,' I said, wiping some chicken muck from my sleeve.

The vet gestured at her box of surgical equipment. 'Would you care to observe? We're castrating Mr Whyman's little minipig.'

Immediately, this felt like the wrong thing to say to my neighbour. Somehow she had made 'minipig' sound like a pet name for my genitals.

'I'm sure Roddie has better things to do,' I said, chuckling nervously. 'And I can assure you there won't be any more noise.' I stopped there, only because Butch chose that moment to fall under the spell of the tranquilliser. He might've been smaller than his sister, but he was dense. Toppling into the wallow, the splash he created was enough to crash across our feet. Sombrely, I returned my attention to my neighbour. 'If you'll excuse us,' I said, 'we have work to do.'

By the vet's estimation, we had half an hour before Butch returned to his senses. The operation would be performed outside the pigsty, on what was left of the grass. Watching Roxi's attempt to revive her brother made me realise this was a precautionary move. It was touching to see her try in vain to shovel some life back into him. At the same time, I realised it was the equivalent of having an over-emotional relative in the operating theatre. One prone to trashing equipment or wolfing down medication without first removing it from the bottle.

Butch was small enough for the vet to carry out without my assistance. As she cradled him in her arms, it was strange to see this animal up close and so still. For the first time, I realised just how deep his wrinkles ran. With long, grey whiskers sprouting from folds around his snout, he looked like a beast that had been around for centuries, not a matter of months.

I followed the vet, carrying her kit box, and closed the gate behind us. Roxi looked lost without Butch, but only for a moment. As ever, her sense of smell took over. Within seconds, she was on a hunt for any last shards of eggshell from her earlier plundering. Behind me, the vet had just laid Butch on his belly. I turned to find her arranging his back legs so they splayed behind him, and offered to make us both a cup of tea.

'No need,' she said, arranging a cloth beside him for her instruments. 'I am happy for you to stay and watch.'

I had hoped she would let me go. Instead, I found myself with an invitation I couldn't turn down. Not without looking feeble or plain rude. As she donned a pair of surgical gloves, I even considered whether the offer of a glass of wine might prove more tempting to her. Any excuse for me to head inside while the crucial cuts were made. When the vet prepared another syringe, presumably containing a local anaesthetic, I figured a little conversation might stop me from fainting.

'So,' I said, 'have you ever operated on a minipig before?'

'Many times.' She sunk the tip of the needle into Butch's groin. 'But only on cadavers at medical school.' Immediately, I wished I hadn't asked. With the shot delivered, she looked up at me. 'I'm talking about the laboratory breed of minipig. The *proper* sort. They still contain the same organs as a standard pig. It's just they take up less space in the freezers.'

Verbally and visually, this was too much information. As she returned her attention to Butch, I braced myself for the possibility that the next step would involve scalpels and bloodshed. Instead, she rummaged in her case and produced an aerosol can. Shaking it first, she then sprayed Butch's testicles with what looked like blue paint.

'What's that for?' I asked.

'Comedy effect,' replied the vet, appraising her work.

I glanced at Roxi and the chickens, as if somehow they would confirm that what we had here was a crazy lady.

'Really?'

'Of course not.' She rolled her eyes and then flashed me a smile. 'It's an antiseptic.'

I chuckled nervously, and then recoiled in silence as she picked up a scalpel and seemed to size up Butch with it.

'How long will this take?' I asked, with my back turned to her. 'I'm sorry but I really can't witness this bit.'

'Your loss. I just need to apply some clamps and then we'll be on our way.' She added nothing more for a short while. Nevertheless, each crunching sound that followed caused me to grimace in different ways. 'OK. Everything is in place. The next step is really very simple . . . one . . . two . . . *there*!'

I am not sure what persuaded me to turn around. She just sounded so elated, like a midwife at the moment of a birth. Basically, what I found myself looking at was a beautiful woman with Butch's detached gonads glistening in the palm of her hand. Smiling in a way that I could only describe as enticing, the vet invited me to hold them.

'I'll pass,' I said quickly, and raised my arms in surrender. 'To be frank with you, this is going to mess up my dreams for quite a while.'

'Some even say they taste good,' she added. The way the vet was bouncing them made me think she was about to demonstrate some kind of one-handed juggling trick. 'It isn't every day you get an opportunity like this.'

I just couldn't be sure if she was projecting a genuine enthusiasm for veterinary science or indulging in some weird brand of humiliation. All I know is that I had never been so relieved to hear my mobile phone ring. The fact that one of the kids must have played with the ringtones didn't bother me at all. It took me a moment to pull it from my pocket and stop it from oinking. Then I seized the opportunity to excuse myself so that I could answer it.

'It's me,' said Emma. 'How's it going?'

'Good!' I glanced around to see if I was being eavesdropped on. Mercifully, the vet had returned her attention to the sedated minipig. 'Come home,' I pleaded in a whisper. 'Help me!'

'What's happened?' Emma sounded concerned. 'Is Butch OK?'

'It's not him you should be worried about,' I whispered. 'Your vet is bonkers!'

'Well, she sounded very nice on the phone.'

'She's just been tormenting me with Butch's testicles,' I said, retreating further up the garden. 'I need you here.'

Down the line, I could hear her cup the mouthpiece and confer with a colleague.

'I'm late for a meeting,' she said. 'I just wanted to remind you to ask about worming both minipigs. The lady from Trading Standards said we had to do it within the month.'

'Emma, I—'

As the phone went dead, I turned to find the vet was now focused on applying a couple of stitches to Butch's nether regions. Thankfully, his now redundant testes had been sealed inside what looked like a plastic sandwich bag.

'He's going to be sore for a few days,' she said, pulling the thread through. 'Although I suppose that's stating the obvious.'

If Butch's experience was anything like mine, I thought to

myself, he could at least expect an extended period on the sofa and a steady supply of tea and biscuits.

'That was my wife,' I said.

'I know,' she replied, without further explanation.

I drew breath and tried again. 'She wants me to ask if you can worm Butch and Roxi as well.'

The vet tied off the thread. 'I can,' she said, 'but you'd be the first pig keeper not to do it yourself.'

'Oh, right. Is it easy, then?'

'Once you know how,' she said. 'It's also far cheaper than getting me to worm them.'

I liked the sound of this. 'So, where do I get hold of the stuff?'

'The wormer? I have some in the car.' She began to fold her instruments away. 'I would just wait until your boy here has recovered. Give him a couple of days and you can give him the injection.'

I had forgotten that the Trading Standards lady had told me it came down to two methods.

'Do you have it in feed form?' I asked. 'We only need a small amount.'

The vet placed a finger on Butch's neck, either deaf to my request or ignoring it completely. 'This is the spot. Same with the female.'

I waited for her to look up at me.

'Please,' I said, appealing to her better nature. 'You must keep some pellets in the car?'

Smiling to herself, the vet removed a tray from the box beside her. She plucked out a syringe and placed a little medicine bottle on the grass beside it.

'Be a brave boy,' she told me, 'like your little pig here.' On the grass, Butch remained quite motionless. Unlike Roxi, who had returned to persecuting the chickens. This time, she had lifted her front trotters onto the henhouse, presumably in a bid to tip it over. Inside the coop, the four ex-batts were falling over one another in a bid to peck at her through the wire. Having had her attention drawn by the sound of vicious clucking, the vet looked back at me.

'The best way to keep the pigs away from the eggs is to raise that henhouse off the ground. You should probably do it as a matter of urgency, in fact. You might be losing eggs, but if they continue to fight your pigs could lose an eye.'

'That sounds serious,' I said. 'I hadn't considered that.'

'If you know anyone who's good at DIY,' she said, with just a glance at the garden junk I had used to build a barricade, 'now might be the time to contact them.'

Reluctantly, I told her I knew just the man. 'I've been meaning to ask him about building a bigger ark. There's plenty of room for Butch, but his sister can barely squeeze inside now.'

The vet looked at me quizzically. 'His *sister*? Mr Whyman, these minipigs aren't siblings. No way. Just look at their markings, not to mention the difference in size.'

'But they came from the same litter!' I protested. 'It was the pictures of the piglets together that made my wife go weak at the knees.'

The vet considered this. 'Did you pay much for them?' she asked. 'Actually, don't answer that. I think I can guess.'

'So, what are you suggesting? That we've been conned?'

She thought about this. 'I expect your wife loved the idea of siblings, no? A little brother and sister? That's got to be worth paying out a few extra pennies for.'

I looked at the ground and nodded. 'Could this get any worse?' I muttered.

'Don't get me wrong,' she added. 'They're both fine-looking, healthy specimens, and there's a growing demand for so-called minipigs.' She paused there, and looked sadly at Butch. 'It's a shame you can't breed them now.'

20

Housebound

On Emma's return from work, I had two items of news to break to her. Both were less than great. I had thought hard about the order of importance. I also figured it would be best to begin in the kitchen, sitting down with a glass of wine before heading down to the pigsty. What put paid to my plans was the kids. I should've known it's psychologically impossible for anyone under the age of 14 to keep a secret.

'The minipigs aren't related,' declared Lou, the first to reach her mother as she stepped inside the house. 'And you need to inject them. With needles. If you get it wrong they could probably die. In agony.'

I had come in at the tail end of this rather too full disclosure. Standing behind our eldest daughter, with Emma looking at me, I drew breath to explain.

'*Mum*!' This was May, appealing for Emma's attention from the front room. 'Dad's built something upsetting! Tell him it has to come down!'

Chuckling nervously, I told May to give her mother some space. The little ones had been drawn by her return. They were clamouring for a hug, and secretly some little present. 'Let her come in and kick off her heels. Glass of red or white?' I offered.

Emma seemed lost in thought, clearly spooling through

everything she had just been told. 'I need to see the minipigs,' she said after a moment, sounding a little panicked.

And so it was that despite my preparations, four children and I followed Emma to the pigsty. That she broke into a run on reaching the grass didn't bode well.

'Butch is doing fine!' I shouted after her, and batted away the last of the day's flies. 'Take it from me, he'll need a day or so for his back legs to stop looking so bandy, but he's guaranteed to make a full recovery.'

Emma stopped before the pigsty. Understandably, only Roxi ventured forward to greet her. Butch remained inside the ark. His head was visible in the doorway, resting in the late afternoon sun. Crouching at the fence, Emma petted the minipig as if she had been away from her for a decade.

'It's OK now,' she could be heard to say. 'I'm home. No more bad things can happen.'

'She's perfectly fine,' I said in protest. 'They both are.'

Emma faced me. 'What? In the shadow of *that* thing?'

I turned my attention to the structure in question. It had taken me much of the afternoon to assemble a temporary solution to the egg-plundering problem that didn't just fall apart. It only had to hold up until Tom could find the time to drop by. This time, following the vet's advice, there was no way that Roxi could gain access. Not unless she learned to climb a steep ladder built from two lengths of trellis. The chickens seemed quite content perched at different levels, even if that did just draw attention to how squeamishly featherless they looked.

As a framework for my creation, the henhouse itself sat upon the top of what was once a wire puppy cage.

For her first year, we had been advised to house Sesi in it at night. We were told this was to help her feel secure and also prevent her from tearing up the house. I think the trainer must have had my fear of prison in mind, too. This was because the cage also stopped her from slaughtering anyone foolish enough to break in and then leaving me to answer the charges for possession of a demon dog. We had used it for about eight months, before Sesi grew so big that it looked more like we were raising a battery hound. By then, with the child gates in place, it turned into a room in the house that nobody wanted to spend time in on account of the hair and the mud. I removed the cage and gave her a cushioned mat, which quickly became filthy thanks to daily walks in the wood. At times, when I had taken her out, any visitor who stepped into our kitchen would have been forgiven for thinking that we were a nice, middle-class family who also happened to be housing a hostage. As for the cage, once folded flat it joined the rest of the junk in the shed, filling up space that would otherwise be taken by a workbench and a tool shelf. I really hadn't thought I would ever find another use for it, until now.

Having reassembled the cage and pinned it in place with all the children's plastic windmills I could find, I genuinely thought that Emma would appreciate that this was a work of ingenuity in place of craftsmanship. I had even taken steps to deter the minipigs from using brute force to reach their prize. This I had achieved by slotting two folded deckchairs between the top of the cage and the henhouse. They stuck out like wings, but would stop the minipigs from rising up to shove the house to the ground. I had used the kite string again, just to make sure that everything was bound tight. Instead of recognising how resourceful I had been, it seemed to me that my wife was very much on the side of May and Lou.

'Call Tom,' she instructed me, matter of factly. 'Tell him we need his help as a matter of urgency.'

'But why?' I protested, as the little ones let themselves into the pigsty. 'Butch and Roxi can't get up there and the chickens have a safe place to roost. It works perfectly.'

'But it *looks* terrible! It's like you're trying to build a mutant robot in our back garden!'

'It's no worse than my last effort,' I pointed out, before asking Honey and Frank to stop shaking the cage. 'If you do that,' I said, 'it *will* fall off. Come on, guys. Use your brains.'

'More immediately,' she added, 'it's unsafe.'

'Only for small children,' I said. 'The minipigs can't knock it over. They're not big enough. Now, I agree it doesn't look great, and yes it can be seen from the lane, but it's only a stopgap measure, so what's the big deal?'

Emma sighed deeply. I just knew this wasn't only about the fact that I was responsible for an eyesore. We had other issues to address as well. In the pigsty, Honey and Frank were petting Roxi. Butch just wasn't interested in anything but lying there looking sorry for himself. I knew that feeling well.

'So,' said Emma next, with Lou and May at her side once again. 'Who told you the minipigs weren't related? Was it the vet? On the phone, you told me she was a crazy.'

'She took her clothes off in front of Dad,' said Lou, narrowing her eyes at me. 'He told me all about it. He said she got naked in the driveway.'

I pinched the bridge of my nose. My headache was coming back. 'She didn't undress for my benefit exactly.'

'But nevertheless she undressed in our drive?' Emma looked set to shield the children from the rest of this exchange. 'So whose benefit was it for?'

Patiently, I explained what had happened. 'It was just one of those things,' I assured her. 'I didn't make a big deal out of it.'

'Yes, you did!' Again, Lou took me to task. It wasn't hard to work out why. Had I agreed to remove the dog cage, she'd have been the first to defend me. Instead, with her eyes glittering, my eldest daughter set about making life difficult. 'The first thing you told us when we got back from school was that you'd seen a lady vet with no clothes on. You made it sound like the highpoint of your year.'

For a man who had defaulted to looking after a small army of children and pets every working day, it didn't seem so unreasonable. Still, I knew better than to fight my corner with that claim.

'It was just a silly thing that happened,' I said instead, appealing to Emma as much as Lou and May. 'I thought sharing it would make you smile, OK? Oh, come on. It isn't every day that a beautiful woman strips down to her lingerie in front of me . . . wait a minute, I didn't mean it like that. The vet wasn't even aware that I was standing behind her . . . not that I set out to watch.' I stopped there, unable to trust my inner defence barrister to get me off the hook. 'Anyway,' I said, anxious to move on, 'it was her professional opinion that Butch and Roxi cannot be related. It was kind of obvious to me as soon as she said it. Just look at them.'

In silence, apart from the little ones, my family contemplated the minipigs. After a moment, Emma took her BlackBerry in hand.

'There's only one way to settle this,' she said, jabbing at the keys. 'I can get the breeder's number from her website.'

As she waited for the page to load, Roxi circled behind Honey and Frank in a bid to seek out the best tickle. Startlingly, I realised she was now waist high to the little ones. I glanced at her former brother. Quite clearly, she and Butch shared just one thing, which was an ark that could no longer house them comfortably. Just then, Emma tutted and cursed under her breath.

'What's up?' I asked.

She showed me the screen. 'The breeder's website is down,' she replied. 'It looks terminal, too.'

'Let's not jump to conclusions,' I said. 'There may be a perfectly innocent reason why we can't track it down.'

While Emma checked the connection once more, I noticed Lou was fretting with her hands. 'They are still minipigs, aren't they? I've told everyone now.'

'Of course they're minipigs,' I assured her, and glanced anxiously at Roxi.

'Minipigs with worms,' observed May, who promptly questioned whether Frank and Honey should be in such close quarters. 'Dad, you need to bath the little ones as soon as they go back in the house. And get rid of their clothes. Burn them or something.'

'That would just be a waste of good clothes,' I assured May.

'A boil wash at the very least. We don't want parasites in the house. Someone has to think of Miso.'

'Relax,' I said. 'The vet advised me we could wait until Butch was feeling a bit better about himself before we put him through any more stress.'

Looking tense and unhappy, Emma turned and called the little ones in from the pigsty.

'It was a job for the vet,' she muttered.

'At a cost,' I pointed out. 'She's given me everything you need to get the job done.'

'Me?' said Emma in surprise. 'Why me?'

I had hoped that she wouldn't pick me up on this. As Butch's castration had reminded me how much I hated needles, my plan was to subtly suggest it was her duty to administer the wormer. Unfortunately, I had forgotten that my wife spent her professional life scouring documents and contracts for clauses that might've been slipped in under the radar.

'Very well,' I said in resignation. 'I'll ask Tom to do it.'

It was getting late. With the sun setting, the chickens were beginning to climb towards the henhouse. Family tension wasn't a good way to wind down the little ones, and so I suggested that we headed inside. As I ushered Frank and Honey towards the back yard, and the promise of a bubble bath, Lou and May continued to grumble behind me about the state of the pigsty. I turned to reprimand them, only to receive a caution myself.

'You will *not* ask Tom,' said Emma, who had yet to follow us. 'That poor man can't be expected to do every job that's too much for you. If you're not up to worming the minipigs, I'll just have to do it myself.'

Like the close of a tense business meeting, nothing was said about the central issues for some time.

That evening, Emma and I faced the task of getting the little ones to bed. While they called upon an arsenal of resistance tactics, May and Lou plugged themselves into the laptop downstairs and fired up the webcam. We had house rules for this kind of thing. If the older girls wanted to communicate with their friends in this way, it had to be done in a space where we could keep an eye on them. Despite the presence of minipigs in our lives, we considered ourselves to be responsible adults. The only problem with this arrangement was that the webcam's field of vision effectively became a No Go Zone.

I had learned this to my cost once. Having sauntered into the kitchen to prepare supper, I tied on Emma's apron to protect my top. I had given it to her one Christmas. It featured a life-size but headless print of a Burlesque dancer. The response from Lou and May was uncalled for, I felt. I was clothed, after all, and had resisted the temptation to sweep behind them with jazz hands and a tap dance. Even so, from that moment on, whenever the webcam was on we were allowed to prepare food in the background as long as we obeyed two non-negotiable but wholly unreasonable rules.

Rule number one prevented Emma or myself from turning to face the webcam. Rule number two insisted that we could cook anything we liked, so long as it didn't involve the use of supermarket value brands.

For supper that night, I opted to drive out and pick up a take away. As we lived in the middle of nowhere, this involved a twenty-mile round trip. Given the atmosphere in the house however, I saw it as a chance to find some space I could call my own.

The four of us ate in silence. Lou and May were tuned into a fashion makeover show on the television, while Emma leafed through a brochure in front of her plate. I watched the programme for a little bit. After a while, I wondered whether it would be more entertaining to gouge out my eye with a chopstick.

'What are you reading?' I asked Emma instead.

She turned a page, chewing on dim sum. 'Just a holiday brochure,' she said. 'We haven't had a break this year. I think we need one.'

Judging by the pictures of sun-kissed, palm-fringed beaches, the destination she had in mind was located several time zones beyond our budget. More importantly, Emma seemed to have forgotten about our responsibilities closer to home.

'What about the minipigs?' I asked, picking at my food now. 'They'd trash the beach.'

'Don't be flippant,' she replied. 'They'll stay here.'

'Home alone?'

'Of course not. We'll find someone.'

I chewed this over for a moment. 'Someone like who?'

'I don't know. Tom, perhaps?'

'You can't ask him to take care of Butch and Roxi,' I said, mindful of her decision not to call upon his help every time we found ourselves in a fix. 'For one thing, he'd have to be here at the crack of dawn to stop them screaming to be fed.'

'But he keeps pigs,' she replied. 'When does he feed them?'

'In his own time,' I said. 'His pigs know who's in charge.'

Ignoring my comment, Emma went back to perusing the brochure.

'How about kennels?' she suggested, without looking up.

I had just collected some noodles with my chopsticks when she said this. They didn't make it to my mouth.

'Kennels? For minipigs? Is there a demand for that sort of thing?'

'Of course,' said Lou, still watching the television. 'From the celebrities. They're always jetting around. Who do you think looks after their minipigs?'

'People.' Like her sister, May spoke with her eyes glued to the screen. 'Celebrities have people to do that sort of thing. We don't have people. We can't afford to hire anyone. We just have pets. Lots of pets. And we're not going anywhere because Miso would freak out.'

'At least freaking out would be a sign of life,' I muttered under my breath, before addressing Emma directly.

'The only kennels around here are for dogs. There isn't one I know that would take on the minipigs,' I said. 'The paperwork would be impossible for one thing. We have to face facts here. While our family includes Butch and Roxi, we just can't go on holiday. We'd need to pay for a house sitter, and we don't have the funds for that. All our money goes on keeping all our children and animals sheltered, fed and watered.'

This time Emma faced me. She looked at me like I was being ridiculous. 'Butch and Roxi could live for fifteen years. Are you saying we're housebound for the next decade and a half?'

In response, out of frustration more than anything else, I pushed my plate away. The scraping sound drew Lou from the television once more. She flicked her attention from one parent to the other, and then asked a question with the kind of timing that only a teenager could possess.

'So,' she said, 'does that mean it's still OK for me to have a sleepover here tomorrow night? I was hoping I could hang out in the front room with the minipigs and my friends. You wouldn't mind spending the evening in the kitchen, would you?'

Moments after I had finished with my plate, Emma seemed to lose her appetite, too.

'Go ahead,' she said. 'Consider it a holiday.'

They say you should never go to bed on an argument. In our case, turning in with tension hanging over us would inevitably mean waking the next morning to find a dark thundercloud over my wife's side.

Emma didn't forget, unlike me after a glass and a half of wine. So, in a bid to clear the air, I called Tom after supper. Firstly, I asked if he might be able to help me create some kind of minipig defence system around the henhouse. Tom sounded amused when I told him about the issue with the eggs, but assured me it would not be a problem. Next, I explained that Butch and Roxi no longer had room in their sleeping quarters in which to swing a cat. Privately, I wished that someone could try it out with Miso, if only in the hope that it would knock some life back into him.

When Tom began to chuckle outright, I told him it was no laughing matter. In response, he informed me that Emma had just called, and put in exactly the same request. I smiled to myself and said that I looked forward to catching up with him.

That night, in bed, Emma and I read until very late. I chose a manual about chicken keeping. Emma pushed on with a novel about a woman who fakes her own death. The turning of every page or so was marked by a thumping sound from outside. Most of the time, the noise just made our window vibrate. Every now and then, it was loud enough to set the dog off downstairs. Sleep just was not an option. Eventually, Emma rested her book on the duvet.

'Can I ask you something?' she said. 'Are you cross with me?'

'Cross?' I stopped reading for a moment. 'I'm just tired,' I admitted, staring at the ceiling. 'I feel like I'm a new parent all over again. It's like that stage of my life is never going to end.'

My response was met by silence. Finally, Emma turned on her pillow to face me. 'I'm beginning to think we've bitten off more than we can chew. This evening has really brought that home to me.'

'Are you still fretting about holidays?' I asked her, though I couldn't offer a solution that would make her feel better.

Nor did Emma have anything more to add.

After a moment, I reached up and switched off the light. Outside, Roxi turned in the ark. It took her a long while to settle, half an hour, perhaps. Fortunately it no longer came to Sesi's attention. Even so, as Emma's breathing told me she was slipping into a slumber, I found myself lying there with minipigs on my mind. I didn't think that Emma was having serious second thoughts about taking on Butch and Roxi. Despite her drive to turn our family into a farmstead, she had never given up on an animal, no matter how challenging they proved. My problem was it felt like I kept having to make the sacrifices to accommodate them, something that was starting to make me feel squeezed. I had yet to form the kind of bond with Butch and Roxi that Emma clearly enjoyed. I just hoped that in time Tom's advice would come true, and I would see through the workload to something that was basically good for the soul.

What wasn't beneficial to my life, or any prospect of sleep, was the sound of hushed whispering on the lane outside the garden.

I'm not sure whether I had nodded off beforehand, but when I heard it I snapped my eyes open. Pre-minipigs, the only thing to wake us was the bark of a fox from the woods. What I could hear was unmistakeable, but the minipigs weren't the cause of it. It was enough for me to draw back the covers and creep to the window. I parted the curtains by a fraction, and peered out into the lane. Outside, under moonlight, two men were standing beside the fence. I could hear them talking quietly to one another. I didn't recognise either of them. One was heavy-set, the other skinny with a baseball cap pulled low over his face. It wasn't unusual for people walking by our house to be drawn to look over. You don't hear pigs honking from a back garden every day, after all. Only this was the middle of the night, and Butch and Roxi were fast asleep inside the ark. For a moment I considered reaching under the bed for the plastic light sabre. Then I realised my dressing gown was in the wash. Somehow, I didn't think this pair would be too rattled if confronted by a naked Jedi.

The heavy-set figure chuckled just then, only to be shushed by his friend, which he didn't take well. As I watched, it crossed my mind that they could of course just be drunks making their way back from the village pub. They would have come quite a distance, but it seemed a likely explanation. Nevertheless, I wasn't happy about them squabbling outside our house, even if they were just doing so in a whisper. If they picked up on the presence of minipigs, and word spread back to the pub, we'd have a succession of pickled farmhands wobbling their way up the hill just to see for themselves.

'Haven't you got homes to go to?' I muttered, feeling irritated now by their presence. 'Come on, guys. There's nothing to see here.'

Emma was asleep. She would only be alarmed if I woke her up by switching on the bedroom light in a bid to encourage the pair on their way. Then she'd stew on it and suggest bringing Butch and Roxi back inside. I certainly didn't want minipigs in the house again, but then nor was I keen on them becoming a late-night attraction for inebriates. To my relief, Sesi must've heard them just then. Had I been in their shoes, I too would've been persuaded to hurry away by what sounded like a human shapeshifting into a werewolf.

21

On Space

It was a relief to have work to do outside the next day. As a steady stream of Lou's friends arrived for the sleepover, so it felt like the oxygen in the house was thinning.

I had assumed Tom would show up with materials to extend the ark. I'm not sure why I thought this was what he would be doing. When I asked him, he just invited me to take a good look at the existing structure.

'What did you have in mind?' He stood behind the ark and spread his hands. 'A glass sunroom, perhaps? A second storey with skylights?'

'Obviously not,' I replied, in no mood for being teased about the issue. 'But can't you just make it bigger somehow?'

'No need.' Tom gestured at the shed. 'Not when you have the perfect housing in place already.'

'What?' At first I thought he was joking. 'That's my shed!'

'I'm thinking we just cut a little doorway into the side there, section off the back of the interior, and your minipigs will have themselves all the room they need.'

I considered Tom's proposal. It wasn't like I used the shed for anything but storage, but still I considered it to be a space I claimed as my own. Nonetheless, it was a solution. One that still allowed me some access rights. 'How much room do you plan to give them?'

'Dunno,' he said with a shrug. 'I guess about the size of a chest freezer.'

I dwelled on this for a moment.

'How about we just fit a real chest freezer,' I suggested. 'It might be the answer to a lot of my problems.'

'You don't mean that.' Tom appraised the shed again. 'It would be a shame to let a fine building like this go to waste.'

'But do you think it'll work?' I asked. 'And what about the chickens?'

Tom crossed the pigsty to the henhouse. It was still sitting on top of the dog cage with the trellis ladder attached. He rocked the cage back and forth. The trellis creaked and one of the deckchairs slipped out of position. 'I think we need to make sure that both teams have a safe place they can call home,' he observed. 'I can fix this, too.'

'At least my efforts are protecting the eggs.' I stepped around to open up the henhouse port. Inside, to my surprise, I found the nest was empty. Even the chickens were absent, I realised. I looked around. Butch was watching us from the ark, having hobbled out briefly when we first showed up. Only then did I think it was unusual that Roxi hadn't ventured out as well. I could hear her in the ark, brushing up against the sloped walls, and that's when my heart missed a beat.

'Oh, no!' For one horrible moment, I feared her taste for eggs had extended to the producers. 'Tom, help me open it up.'

In order to access the ark to clean it, Tom had hinged one side panel. Without time to explain myself, but sensing my alarm, he helped me lift the panel so it looked more like an awning. Together, we crouched to look inside. Tom glanced at me with some pity in his eye.

'Seems your flock didn't need your help after all,' he said.

Inside, four hens were nesting contentedly in the straw

beside Roxi. They had laid their eggs at the back, not that there was anything left but shell fragments. Roxi had taken care of the contents. In fact, she was still licking yolk from her chops. I looked at Tom, thinking through some explanation.

'Unless I'm mistaken,' I said, 'it looks like they've struck a deal.'

'Seems so.' Tom gestured for me to close down the side and leave them in peace. 'The chickens get protection from predators in there, plus a little warmth.'

'While Roxi gets to stuff herself on all the eggs she can eat.' I stepped back from the ark, feeling a little cheesed off. 'Can we put a stop to this? I didn't just take on some battery hens to provide them with a better life. I was expecting them to give me something back in return.'

Tom chewed on his lip. 'Unless you want to abandon your plan to keep them together in the pigsty, there isn't much you can do.'

'The original plan was for Butch and Roxi to deter the fox from eating the chickens. This is just wrong.'

'It's like those birds that clean crocodile's teeth,' he observed, 'only this relationship is between minipigs and poultry.'

I tutted and kicked at the dirt. This wasn't an arrangement I had anticipated, but at least it meant my hens were safe from harm.

'If they're going to live under the same roof,' I said, with my shed in mind, 'I suppose that's all the more reason to give them the space you suggest.'

'Then let's go to work.' Tom had brought his tools in a leather carrier. As he set off to fetch them from his Land Rover, I returned to the house to find mine. They were in the kitchen, comprising of a kettle, two mugs, some milk and teabags.

I made a stronger brew than usual. This wasn't just to suit

Tom's taste. I needed something to help me feel a bit more alert. The truth was I felt tired. Tired and a little troubled. The night before, after the two men had made off down the lane, I lay in bed and worried. Even if they had just been innocent passers-by, I didn't feel comfortable with Butch and Roxi being on such public view. I told myself that I was overreacting. I also knew the reason why. Having lost loved ones, and been powerless to do anything about it, I was basically sensitive to any kind of threat to my family. Worries always seemed much worse in darkness, of course, while the sound of Roxi turning every now and then didn't help me get to sleep.

When I did drop off, all I did was dream that the garden got flooded. It was so deep that I couldn't make it down to the bottom of the garden in time to save the minipigs from sharks.

Clearing the shed should have been therapeutic. I was chucking out junk, and creating much-needed space. A lot of it I had already used in my failed attempt to protect the chickens in the enclosure. That now lay in a heap in the garden, thrown out unceremoniously by Tom. As I added to the pile, Lou came down to escort Butch and Roxi into the house. They were only going in for the early stages of a sleepover, but I had to admit to myself that it did at least mean they were out of harm's way. I must have looked preoccupied, because eventually Tom stopped sawing out timber on his workhorse and asked me outright.

'What's on your mind? Is it the shed? I realise it means a lot to a man. It's his cave, isn't it? But you're not going to be shedless, Matt. It's just going to be . . . smaller.'

I listened to what Tom said, and then dumped a stack of plastic plant pots onto the grass.

'I'll survive,' I told him. 'It isn't the first sacrifice I've made for these minipigs.'

Tom placed one foot on the workhorse. Resting on his elbows, with the hens at his feet, he looked me in the eyes.

'I've told you there's a joy to be had from pig keeping, but you don't seem to be feeling it.'

'It's not that,' I said. 'I'm even getting used to the flies.'

Tom looked at the concrete floor for a moment. 'Pardon me for asking, but is everything alright at home? When Emma called me last night, she seemed a little down about you.'

'We'll be fine,' I said. 'In different ways, we're both just having a wobble about the wisdom of bringing these things into our lives. I hear what you're saying about the joys, but you have the space. All they've done is make me feel anxious.' I stopped there, and considered the fence. 'We had a couple of guys looking in from there late last night. It was probably nothing but I didn't like it one bit.'

Without a word, Tom listened to my account of what I had seen from the window.

'I can't say I'm surprised,' he said. 'These days it seems like everyone wants a minipig.'

'What would you do?' I asked. 'In my shoes?'

Without hesitation, Tom said, 'Kill them.'

'Eh? All they did was check out the pigsty.'

'Not them,' said Tom. 'The minipigs.'

For a second I was lost for words. His clarification made it sound even worse.

'How can you say that?' I asked. 'They're a pain, for sure, but I can't just have them put down.'

Tom shrugged. 'This morning I was telling my missus about all the grief they were causing you and she asked why you didn't just take them to slaughter. Along with a quiet life, at least you'd get a couple of meals out of them.'

I looked at him in disbelief. 'We don't talk about cats or dogs in that way. When Sesi starts to get on in life and the walks become a chore, do you expect to see me look at her in a different way and start looking up some recipes? No, you don't! It would be outrageous and distasteful.'

Briefly, Tom looked taken aback. 'Something tells me these minipigs are here to stay,' he said.

I drew breath to suggest otherwise, but he was right. It wasn't just the concrete under Tom's feet. Butch and Roxi had found their place, not just in the shed but in the hearts of Emma and the kids.

'It's important they stay safe,' I said instead. 'For everyone's sake.'

Tom ran his gaze around the perimeter of the pigsty. 'We could always wire up an electric fence if it means you'll sleep at night.' He seemed to consider the idea for a moment, only to drop it on spotting the two blue circles on the grass. 'What's that?'

I explained that it marked the spot where Butch had been castrated.

'Not that it turned out to be necessary,' I added. 'The vet says it's genetically impossible that they can be brother and sister. The trouble is she only told me after Butch's little testicles had been sealed inside a bag.'

Tom's brow furrowed. 'You weren't planning on breeding them, were you?'

'Of course not!' I scoffed at the suggestion. 'Two minipigs give me enough grief. Can you imagine the hell we would create by bringing more into the world?'

I expected Tom to agree with me. Instead he seemed a little lost in thought.

'There's money in minipigs,' he said with a shrug, and then returned to work.

Looking back, having one child was a bit of a part-time position. Lou was a central aspect of our lives, but if I needed time to myself then I could just hand her to Emma, and she could do likewise when she needed space. Two children was a commitment. All of sudden, we both had our hands full. This was most apparent to us at weekends, previously a time to relax and recharge. If anything, without the kids at nursery or the sanctuary of work, those days were more demanding. They were still fun, a chance for us all to be together, but secretly I think we both viewed Monday mornings as an opportunity to unwind. With four kids in the fold, both Saturdays and Sundays had become a cliff face of commitments. For us, weekends involved juggling the demands of the children with grocery shopping and housework. That afternoon, for example, as Tom toiled on the shed conversion and Lou held court with the minipigs and her friends, it fell to me to take the little ones to swimming lessons. I felt bad about leaving Tom to do all the hard work. When I told him I had to break away for an hour, however, he looked quietly relieved.

'I've been thinking,' he said, having already fitted the internal wall of what would be the minipigs' new home, 'without the junk, there's more room in this shed than I first thought. If you'll just leave me to my own devices for a little while, I'll have it looking like a place you can be proud to spend time in.'

With Frank and Honey in the water, each in a different class, I settled in the spectators' gallery along with all the other mums, dads, grandparents and carers. Some had children with them. Others just tuned out and read the papers. Normally, Emma oversaw the swimming lessons. As Lou and her friends had decided to do some baking, which always seemed to involve carefully sifting icing sugar over every surface, it was decided between her mother and me that I should be out of the house. Just in case I lost what patience I had left and cast a shadow over the sleepover.

It was good to catch up on the little ones' progress. Honey was looking like a natural born swimmer, while Frank, without armbands, no longer threatened to spend more time on the floor of the pool than the surface. It made me realise they were growing up. I just wondered whether Butch and Roxi might mature into quieter minipigs, particularly in the mornings.

I was dwelling on this when the phone in my pocket started oinking. It signalled the arrival of a message from Tom, and came with a picture attachment. Before I'd had a chance to open it, another image arrived. I opened the first one. Evidently, he had finished work on the shed. I was now looking at a square opening in the side wall with a little awning and even a baffle inside to stop the draft from getting in. The second image, showing the interior, was altogether more exciting.

'Tom, we don't deserve you,' I said to myself, and beamed at his handiwork.

I held the phone level, to avoid the glare from the overhead lights, and marvelled at what he had made in my absence. The back of the shed was now host to the minipigs' chest, accessed from the outside and also from a lid on top. Above it, he had fitted a shelf for all the things I might want to store in there; things like tools and lengths of timber. Best of all, Tom had built me a *workbench.* A proper, waist-high station where I could make things.

'You beauty,' I declared, as an attendant whistled from the

opposite side of the pool. 'Oh man, I'm going to be spending quality time with this.'

The whistle sounded again. Dimly, I became aware that one of the attendants was facing the spectators' gallery now, calling up to someone. I looked from my phone to the attendant, and realised she was addressing me directly.

'Sir, taking photographs of minors is forbidden at this pool. Please put your camera phone away.'

I went cold. Already, people around me had turned to see what was going on. Even the swimming instructors had halted their lessons on account of the activity. I heard someone tutting nearby. Another parent placed a protective arm around their small son.

'I'm not doing anything dodgy,' I said, laughing nervously, and showed the attendant my phone. 'You can look if you like,' I added, and wished that I hadn't.

'Sir! If you won't put it away, I'll call security.'

'That won't be necessary,' I said, and pocketed my phone. A mother two rows down was glaring fiercely at me. 'It was just a shed,' I reasoned, as the lessons resumed. 'I keep minipigs in there, not kids.'

The mother muttered something at me that sounded like it would be better suited to a placard outside a courtroom, and faced the pool once more. I glanced at the big clock. Realising I would have to endure another twenty minutes of accusatory looks, I clasped my hands and focused on the little ones in the pool. Then I realised that might be construed as paying too much attention to the children, and looked at the tiles at my feet instead. I even wondered whether I needed to point out that I was here as a father, and not some troubled adult. Eventually, I decided to wait in the reception area until Frank and Honey had finished their lessons. As I moved to leave,

I realised such a retreat was likely to cast me in an even darker light.

'You don't have to worry,' I said in vain, as people withdrew offspring from my path. 'Their mother will be here next week, not me.'

Back home, I had assumed that Emma would share my outrage. If a man was no longer able to sit at the poolside and admire a picture of his shed, just what was the world coming to? As soon as I followed the little ones through the front door, however, I realised more pressing matters needed to be discussed.

'What on earth is going on?'

A mist of icing sugar hung in the hallway, defined by sunbeams through the window. It was such a shock to see, that I didn't register the noise until both minipigs and a gaggle of teenage girls ran from the kitchen to the front room, snorting and bellowing at the top of their voices. It was over in a blink, which is all I could do in their wake.

'Hi!' Emma appeared at the top of the stairs. 'It's quieter up here,' she said. 'Lou and her friends are having a wild time.'

'I can see that,' I said. 'Looks out of control to me.'

'Let them have fun.' Emma stepped aside as the little ones bundled up the steps. 'Butch and Roxi are loving every moment, too.'

As she spoke, the noisy swarm of minipigs and girls crossed back to the kitchen.

'Shouldn't Butch be resting?' I asked. 'I didn't leave the sofa for several days after I handed in my cards.'

'Butch is just being a man about it,' she said with a smile. 'He's even stopped stealing stuff, but I guess that's because he's having so much fun right now.'

I sighed to myself, aware that we still had the sleepover

itself to suffer. I considered reporting what had happened to me at the pool. Instead, in a bid to stay positive, I asked if she had been down to the shed. 'Tom sent me a picture,' I said. 'As well as new sleeping quarters, he's transformed the rest of the space for me. I have a workbench now!'

Emma looked puzzled. 'What do you want one of those for?'

I drew breath to list all the projects I had in mind, only for the stampede to cross the hallway floor once more.

'Never mind,' I said eventually, and elected not to slip off my coat. 'Let's just say you'll always know where to find me from here on out.'

'Not right now,' said Emma. 'Lou's sleepover needs supplies. We need six pizzas, two bottles of fizzy orange and a whole bunch of chocolate, popcorn and sweets.'

'They'll never sleep with that in their system,' I said. 'Think of the additives.'

'It's a treat,' she reminded me. 'Don't spoil the party for them.'

At weekends, once the light went out in the little ones' bedroom, Emma and I would flop on the sofa, turn on the television and switch off from intensive parenting duties. Lou and May were at an age where they no longer needed such close care and attention. They were really quite low maintenance. Except on sleepover nights, of course, when May's stress level needled into the red and her older sister made sure it stayed there.

'Mum! Miso is busting for a wee, but he can't get through the front room to the cat flap because *they're* watching movies!'

May was clutching the cat protectively. The pair had taken refuge in her bedroom, while Emma and I decided to sit it out in the kitchen. The last time I peered into the front room, Butch and Roxi were enjoying belly rubs on the carpet. One of the girls had asked if they could give the minipigs a makeover. This I had overruled on hygiene grounds, as well as the possibility that Roxi would take a shine to lipstick and refuse to let me take it off. Despite all the hassle of having them inside, I did take some comfort from the fact that they now appeared to be fully housetrained. Inside, they really were becoming quite civilised.

I just couldn't say the same for the sleepover girls.

The screams and whooping had subsided after I pleaded with them not to wake the little ones, while the introduction of a romcom on DVD had ensured it stayed that way. Unfortunately, having calmed down volume-wise, they had directed their efforts at scattering popcorn, crisps and sweet wrappers around the front room. Lou was under strict instruction about not feeding Butch and Roxi. Even so, I feared that with so many treats at ground level she would lose control of them and watch helplessly as they ploughed the carpet into strips. With May standing over me now, and Miso looking like he might be better served with a catheter, I figured this was an appropriate time to return the minipigs to the bottom of the garden. There was a long way to go until that most welcome of sights: a parent's car drawing up in the morning, but it would at least bring us one step closer to peace and quiet.

'I haven't had a chance to see where Butch and Roxi are sleeping yet,' said Emma, rising from her chair. 'I'll take care of them.'

'Really?' I watched her place a hand on May's shoulder as they turned for the door. 'Wow!'

Emma paused and looked at me as if she had just missed something.

'"Wow" what?'

I had been taken aback by her offer of help. Not because

she was lazy. That couldn't be further from the truth. But the fact remained that when it came to shutting down all pets for the night, the duty defaulted to me. I just wished I hadn't been quite so vocal in expressing my surprise.

'Wow nothing,' I said, hoping to move off the subject. 'I'll get supper on, shall? There's enough pizza to keep us going for a week.'

Emma's gaze lingered on me.

'It's good there's no TV in here,' she said finally. 'It'll give us a chance to talk.'

She was gone for ages. The pizza only took ten minutes to cook. By the time she returned, bringing Sesi with her via the back door, it was cold. In her absence, the noise had struck up in the front room once more. It was as if Butch and Roxi had possessed the power to pacify teenage girls. From the moment that the minipigs left the party, so Lou's sleepover friends had returned to communicating via the gift of yelling. It had brought the little ones down twice and prompted May to turn up her music to counter the din. The experience had left me feeling frazzled.

'Where have you been?' I asked.

'I took the dog down the lane to stretch her legs,' she replied, and waited until I caught her eye. 'Just pulling my weight.'

I sliced the pizza, aware that she had not forgotten my stupid parting comment.

'What did you think of the shed?' I asked. 'Tom's done a great job, don't you think?'

'Butch and Roxi seem to like it. The chickens, too.'

I realised I had forgotten to tell Emma about the coalition forged in the pigsty. When I explain what had happened, she seemed to miss the point.

'It might mean no more eggs,' I stressed, raising my voice

to be heard over the noise from the front room. 'Which is a shame because that's what keeping chickens is all about.'

Emma sat at the table. 'How about minipigs?' she asked. 'What are they all about?'

I set down the plates, and then decided it was time to open a bottle of wine.

'Butch and Roxi are fine,' I said, struggling to make myself heard. 'It's just they've joined a family on the limit.'

'What does that mean? You keep saying things like this lately.'

I poured her a glass. Upstairs, May had started thumping on the floor in protest. I couldn't hear the television in the front room, such was the noise coming from there. Unless, that is, Lou and her friends had turned it full volume while watching some documentary about teenage exorcism.

'Emma, four children is a lot.'

'Four children is lovely. We're blessed to have them.'

'I'm not complaining about the kids,' I countered. 'I'm just trying to point out that piling the household with an endless procession of animals is running me ragged. It's OK for you. In a way, you can get away from it at work. But my office is also our home. Every time I sit down at my computer, something, somewhere, kicks off and demands my attention.'

'It's never been a problem for you in the past,' said Emma, sounding quite calm despite the racket.

'That's because it wasn't a problem!' I prepared to suggest that the arrival of the minipigs had confirmed what I had feared and pushed things too far. What with all the shouts and shrieking, however, I couldn't hear myself think. 'This is ridiculous.' I leaned my head into the palm of my hand. Just then, the dog started barking. 'The whole household is out of control!'

Emma began to eat as if nothing was out of the ordinary.

'Why can't you just let them have this one night? I think it's great that Lou has so many friends.'

'I'm surprised you haven't offered to adopt them,' I said, and pushed my plate to one side. 'Emma, I'm feeling really squeezed here! Can't you see that?'

She set her fork down, and levelled her gaze at me across the table. 'When I was Lou's age, other people had parties. I wouldn't have dared to ask for a sleepover. It just wouldn't have happened.'

'Maybe your parents had a point,' I replied, and bellowed at the dog to stop barking. I left the table as I did so, feeling a surge of anger and frustration.

'Where are you going now?' Emma demanded to know. 'Don't you dare upset their sleepover!'

'All I ask for is some space to myself!' I snapped. 'And I know just where to find it!'

Emma didn't stop me as I marched from the kitchen. Nor did I hear her come after me as I left the house and slammed the door behind me. It was only when I crept back in to fetch my glass of wine that our eyes met once again.

'Take the bottle,' she said, and shoved it across the table for me to collect. 'It wouldn't do for both of us to run away from our responsibilities.'

22

In the Wilderness

Such was my mood that the shed seemed like the perfect sanctuary. Without Tom's help, I wouldn't even have been able to swing open the door without a mound of outdoor plastic toys spilling out. As it was, with a pale moon shining through the window, I set the wine bottle down on the work-bench and peered through the gloom.

Tom had stored the essential stuff on the shelf, and that included a torch. Fumbling for some garden twine I hadn't realised we owned, I managed to hang it from the roof joist and switch it on. In silence, I looked up and around. Behind me, inside the chest he had built, I heard the sound of straw crackling. I turned to face the minipigs' sleeping quarters. Tom had fitted the chest with a lid because he said it was important to have access in case either of them fell sick. As quietly as I could, I lifted it to peek inside.

There, curled behind the baffle to avoid any draught from outside, Butch and Roxi looked entirely at peace. Both were fast asleep, as were the chickens who occupied the far corner. Just then, as I began to feel a chill in the air, it seemed like the most comfortable place in the world.

'Don't mind me,' I whispered, and closed the lid. 'I'll doss on the floor.'

Closing the door to the shed helped to keep out the cold,

as did several slugs of wine. I could still hear the sleepover in full swing. It just sounded muffled now, and easier to bear. Without my presence, I knew that Emma would just let it go on until Lou and her friends burned themselves out. Reflecting on what had led me to storm from the house, I began to regret the way I had reacted. Having struggled to restore order with minipigs in our lives, the chaos that accompanied the sleepover just pushed me over the edge. Even so, as time began to tick, I wondered how I would save face from this situation. Slumped in the corner, with a shrink-wrapped straw bale as a cushion and draped in a tarpaulin sheet, I nursed the bottle in my hands and doubted somehow that I could claim the moral high ground here. When our bedroom light switched on, I waited for Emma to appear in silhouette and peer through the curtains. I had begun to think that she would break first. If she knocked on the glass, I would have given in with good grace, but it was not to be. After a couple of minutes, usually the amount of time it took her to read a couple of pages of a book before sleep beckoned, the light went off again.

'Great,' I muttered to myself. 'What a marvellous end to my day.'

I was in for the long haul. Unquestionably, Tom had transformed the shed. It just wasn't furnished for an overnight stay. All I could do was find another shrink-wrapped bale, open it up and spread the straw as a makeshift mattress. Then, with enough alcohol in my system to slip into a half slumber, I killed the torch and tucked up under the tarpaulin.

One small mercy was the absence of any noise from Butch and Roxi. At last, they had enough space to turn without waking everyone up. Despite the sound of the sleepover going on inside the house, it meant I nodded off without much trouble.

I might've stayed that way until daybreak, had it not been

for the rumble of an old van. I heard it climbing the lane to the crest, deep into the night, before slowing as it passed our house. If I'd been asleep in the bedroom, it would not have been enough to stir me. As the van drew to a halt just beyond the shed, I was upright within moments. By now, the house was silent, with all the lights off. The engine idled and then stopped. I heard a vehicle door opening, followed by another, and then quiet footsteps on the very edge of Roddie's gravel driveway. Immediately, I thought back to the two men who I had seen in the lane. With the only window in the shed looking out onto the pigsty, all I could do was listen to the pair make their way towards the same spot. As they passed the shed, one of the men coughed. The other one hissed at him to be quiet. A spate of quiet bickering followed. I just knew it was them.

'What do you want with our minipigs?' I whispered to myself.

I patted my pockets, hoping to find my mobile. In a way, it wasn't a problem that I'd left it inside the house. Calling the police from the garden shed and whispering that I was under siege was unlikely to be treated seriously. Instead, as quietly as I could, I rose to my feet. Butch and Roxi did not stir. Some of the chickens clucked drowsily, but these guys weren't interested in four recovering battery hens. This time, the nocturnal activity on the lane didn't rouse my dog. I figured she was wiped out from the stress of the sleepover, and realised I was on my own here. Even with no plastic light sabre to hand, did I jump out and surprise them? I worried that an encounter with a man shacked up in his shed wouldn't exactly serve to scare them away. It was more likely that I'd just look like some headstrong idiot who'd had a row with his wife. Even so, I really felt that I had to do something quickly this time.

With an emergency call out of the question, along with just stepping out to confront them, I opted for a third way. Mostly, this was down to cowardice. Still, as soon as the thought came to mind I committed myself to it. At once, I turned to the chest Tom had built. Opening the lid, I found myself looking down upon two slumbering minipigs. The chickens were settled at the opening; dozing on the job.

'Wake up!' I hissed, and leaned in to prod Roxi in the ribs. 'You too, Butch. Make some noise!'

Startled by my intervention, both minipigs surfaced with a series of rising grunts. It was the hens who were first out onto the concrete. I could see them through the glass. As Butch and Roxi stormed out under the stars, the birds stood around them with their chests puffed out like the poultry equivalent of the presidential secret service. The minipigs, meanwhile, were sniffing the night air inquisitively. With Sesi barking from the house once more, I watched their snouts follow the direction of hurried footfalls retreating down the lane.

The next thing I heard was the van's engine fire up and then tyres grasp for purchase. I should've been relieved when the vehicle promptly pulled away. Instead, as the minipigs retired to their quarters, grumbling in disappointment at not being fed, I settled in the corner once more with a strong suspicion that our visitors would be back. I wasn't frightened, now the van had gone. If anything, I was determined to protect my household, and that included any pets of rising value. Even if it meant adopting a night watch, I decided I would do whatever it took.

Then the cold began to creep in under the tarpaulin, and I decided it might be best to get the electric fencing wired up, just as Tom had suggested.

In the dark, as I struggled to get back to sleep, I considered heading back to the house. Emma would be out for the count, I reasoned. By the time the minipigs heralded the first gleam of sunlight, this nonsense would be forgotten. Despite the early starts, Butch and Roxi made her happy, as did her children. I liked to think her husband figured in that mix as well, even if I wasn't helping my case by sulking in the shed. I began to feel silly about how I had reacted. At the same time, I didn't want to start prowling around in the yard. Not only would it stir the dog, she might also mistake me for a burglar. The prospect of opening the back door, only to be pinned to the flagstones by forty kilos of sinew, tooth and claw, really didn't appeal. Instead, I curled up as best I could, and considered just how I could make amends. I had acted like a fool in flouncing from the house. Nevertheless, this time in the wilderness had served to remind me that I didn't really need the space I thought I craved. I wanted to be with my family, even if we were bursting at the seams.

When it came to me, a solution to our problems, I was on the cusp of drifting off. The way forward, so simple and inspired, had been right in front of my eyes, even though they were shut at the time. In fact, I couldn't be sure that I was dreaming. All the same, I just hoped that I would remember it in the morning.

'Rise and shine, beautiful! It's breakfast time!'

I hadn't expected my wife to wake me, especially not with such affection. What surprised me most of all was how cheerful she sounded. I opened my eyes, feeling cramped and sore, and then took in my surroundings. It was then I figured Emma probably wasn't talking to me. Struggling to my feet, keeping the tarpaulin wrapped around me, I rubbed the condensation from the shed window and peered out. Dressed for the day,

Emma was standing over the minipigs as they wolfed down their food. When she looked up at me, I smiled sheepishly.

'Hi,' I said. 'Looks like it's going to be a nice morning.'

'Let's hope so,' she considered me for a moment. 'Are you going to come out yet? Even if it's a long-term thing, can you spare me five minutes later today to help worm the minipigs?'

I withdrew from the window, discarded the tarpaulin, and opened the door to the shed. Sunlight flooded in. I stepped onto the grass, closed my eyes and bathed in the warmth.

'Nothing like a good night's sleep to help you feel refreshed,' I said, stretching at the same time.

'Absolutely. That was the best eight hours I've had in a long time. For once, your snoring didn't wake me up.' Emma gestured at Butch and Roxi. 'They look as tired as you, though.'

The minipigs had finished eating by now. It took them all of several seconds to finish any food placed under their snouts. The chickens had been standing sentry around Butch and Roxi once more, as if to guarantee an undisturbed meal. With the bowl empty, they duly dispersed to peck at the dirt and await a handful of feed for themselves. I faced Emma once again, and found she was already looking at me.

'About last night,' I began.

'I'm sorry,' she said first. 'Sometimes I forget you're here day and night. But I take that on board, and I'll do what I can to spare you from Butch and Roxi.'

I glanced at the minipigs.

'They're alright really,' I said, mindful of the plan that I had put together overnight. 'But I want to apologise about what I said as much as anything else. I guess the pressure just got to me, but I feel good about things now. A night in the shed has made me aware of a few things. In fact, I have a proposition—'

'Not now,' Emma cut in, before I could finish. 'Lou has guests here.'

'I don't mean *that* kind of proposition.' I grinned at her. 'It's something I think you're going to like.'

Emma brushed some straw from my shoulder. I took this as her way of telling me everything was good between us.

'Whatever it is,' she said, 'can it wait until after we've wormed these minipigs?'

I considered what I had to say. I knew that it would make her day. Still, it made sense to get the chores out of the way first, if only so that Emma would then be free to show me her gratitude. We also had the aftermath of a sleepover to manage. The curtains were shut in the living room still, but as the first car pulled up in the drive I was feeling better about everything. Finally, my longest of nights was drawing to a close.

Like Emma, I found a lot of jobs to do that morning. Even when the last of Lou's friends left, looking in severe need of an additive hit to get them through the day, we continued to tidy and occupy the little ones. I knew what was going on here, and so did she. Eventually, the time would come when we had crossed everything off the list but for one last task.

It was Emma who blinked first.

'We can't put this off any longer,' she said. 'Lou has gone back to bed and May has agreed to watch over the little ones. Let's prove that we can do this.'

She had found me in the office, which was unusual for a Sunday. I had decided it was high time I tried to shore up my desk properly. I was on my back, working underneath it like a mechanic. One who didn't actually own any tools. It meant all I could do was hand tighten those bolts that had come loose, and hope it didn't collapse on top of me in the process.

Easing out from underneath, I found my wife standing over me, hypodermic needle in hand. She looked like a cross between Sharon Stone and Dr Harold Shipman.

'No time like the present,' I said, rising to my feet. 'Let's go find Butch and Roxi.'

She offered me the syringe. I looked at her in surprise. 'I thought we agreed that you would be the one to administer it. Technically, they're your minipigs.'

'I hate needles,' she said.

'So do I.'

Emma narrowed her eyes. 'You've never had an epidural.'

'You've never had an anaesthetic injection in the testicles.'

I reflected on what I had just said, but felt it wasn't necessary to clarify what I meant. Instead, unwilling to get into yet another conflict, I took the syringe from her hand and told her I would do it.

'But you've just said you hate needles.'

'You have to come with me,' I said. 'I'm not doing this on my own.'

Emma's expression softened. 'The meds are in the fridge. I'll go get them.'

'Meds? I hardly think this is going to be a professional operation, but I'll do my best.'

I approached the pigsty with the needle behind my back. I realised Butch and Roxi wouldn't recognise what was about to happen. It just seemed like the sort of stance that a vet might adopt.

'We should fill the chamber now,' suggested Emma. She sounded as nervous and uncertain as I felt. 'If we do it in the pigsty they might smell it and get frightened.'

'Good idea.'

Emma held a small plastic bottle in her hand. She removed

the lid and handed it to me. A foil cap covered the top of the bottle.

'Just plunge the needle through and suck it up,' she said. 'Like on the telly.'

I followed her instruction, drawing five milligrams into the chamber as per the instructions. Once I'd finished, I removed the needle, held it upright and gave it a little squeeze. It was more than enough, in fact, because the whole lot spurted onto the grass.

'Whoops!' I muttered. 'Sensitive, isn't it?'

Emma closed her eyes for a moment. 'What did you even squeeze it for? It's not like we have enough to waste!'

'Because that's what they do, isn't it? On the television.'

'For crying out loud. Give me the needle. I'll fill it.'

With her lips pursed, and using the top of her thumbnail to lift the plunger, she withdrew another measure. I could tell Emma was stressed by what we faced because that's when her patience ebbed away. I looked at the minipigs.

'Who shall we inject first?'

'I think Roxi,' she said. 'If it all goes wrong then at least we won't have killed the cute one.'

I glanced at Emma. 'That's a terrible thing to say.'

'OK, then start with Butch.'

I turned my attention to the little black minipig. He was watching me from behind the wire, his tail twitching from side to side in expectation of a treat. Above all, he liked nothing better than a tickle. This wasn't quite what I had in mind. The vet had advised me to insert the needle just behind the ear. From where I was standing, that looked like a direct line into the brain.

'I feel a bit sick,' I said.

Emma opened the gate to the pigsty, and invited me to step inside. 'Then let's act quickly.'

As soon as I entered, the chickens flocked around us. They were beginning to look less scrawny, with a little more meat on their bones. Even so, there was no denying these birds were working for the minipigs now. They were in the firm. It cost them eggs, but in return they had protection. Having been raised in the poultry equivalent of a permanent penal lockdown, this show of strength was basically a message to their masters: you watch our backs, we'll gang together to watch yours. Stepping cautiously around them, I began to fret about how they might react if they thought I was harming Butch. All of a sudden, the needle looked awfully big.

'These worms,' I said after a moment. 'Can't the minipigs just live with them?'

Emma shook her head. 'It's for the best,' she said. 'You'll be doing a good thing.'

I returned my attention to Butch, and found him gazing up at me. If a dog showed devotion to its master, a minipig offered a look of pure trust. I had the syringe in my hand and total self-loathing in my heart.

'I can't do it,' I admitted, and took a step away. 'I'm sorry.'

Emma looked to the heavens, offered up something I couldn't quite catch, and asked me to hand over the syringe.

'You dithered.' She reached down to pat Butch, and prepared to insert the needle. 'OK, throw him a handful of food to keep him busy. I'm going in.'

I did as she requested. Watching Butch dive upon the pellets, I had to admire her tactics and courage. I felt ill, but as she bent over the minipig and touched the steel tip to his skin, there was clearly no going back.

'That's it,' I said as the needle sunk in all the way. Butch squeaked and flinched, but stayed put. 'Well done!'

I can't say at precisely what moment my wife passed out.

As she depressed the plunger, so her centre of gravity seemed to shift. I watched her rocking over him, wondering what she was doing. Butch worked it out a moment before I did and darted clear. All I could do was scramble to grab the back of her top before her face hit the concrete. Emma must have only been momentarily out for the count. Certainly she was in possession of her senses when I eased her upright once more.

'What are you doing?' she asked, ashen-faced, and then looked at the ground. 'Where's Butch?'

'He's fine,' I said. 'Are you OK? You fainted just then.'

'I did no such thing.' Emma seemed to remember she was holding the syringe just then. Luckily, she had clung onto it when Butch broke away. The downside was that she had scratched her knee with the needle.

'That's going to need some antiseptic on it,' I said.

'I've wormed myself,' she said, as if thinking out loud. 'Bloody hell!'

I helped her to her feet. 'Let me do Roxi,' I said. 'I'm pretty sure I won't see stars.'

'I *didn't* see stars,' Emma insisted. She examined her knee once again. 'Anyway, we can't worm her now. I've contaminated the needle.'

Hiding my sense of relief, I told her I would order another one.

'One wormed minipig is better than none,' I added, in a bid to sound positive.

Emma sighed to herself. She examined her scratch once more.

'At least your wife won't die of lung worm,' she muttered.

I helped Emma to her feet. Some colour had returned to her cheeks, while Butch evidently hadn't come to harm. We might not have been entirely successful in our aims, but I had something to tell her that would more than make up for that.

'You can't die yet,' I told her, seizing the moment. 'I need your help with a breeding programme I have in mind.'

Emma looked puzzled. 'But you can't,' she said. 'You chose to decommission your weapon.'

I had to think about what she meant by this for a moment.

'Not me!' I pointed at Roxi. '*Her*!' This time, Emma took a while to work out what I was getting at. When her expression began to brighten, I just knew this was one winning proposal. 'I need to speak to Tom about the best way to get her pregnant, but it can't be rocket science. What do you think?'

Emma smiled and slipped her arms around me.

'I think that's a *wonderful* idea,' she said. 'I just can't believe it's come from you. All that moaning about not having time to yourself, and now this!'

I pretended to look outraged. In truth, I had my own reasons for making such a dramatic switch in outlook. Firstly, more minipigs would serve to meet my wife's relentless desire to build her brood. A litter would bring me pain, of course, but then who could resist a minipiglet in a teacup? Within no time, all of them would've gone to new homes, and my interest lay in what we'd get in return.

'Minipigs are valuable creatures,' I reminded her. 'We could make enough money from the sale of a litter to *pay* someone to look after Butch and Roxi so we can go on a holiday of your choosing.'

Emma really looked like she was in step with my idea. The prospect of not being able to go away as a family had been a real blow for her. I also held out hope that there would be some spare change from the venture to buy a mountain bike or even a tool belt. As Butch, Roxi and the hens investigated a freshly excavated tree root, Emma had just one thing to clarify.

'We'd want to keep a minipiglet for ourselves, wouldn't we? Maybe even two if it's a big litter.'

This wasn't something I had anticipated. As things were going so well, however, I reminded myself of the profit to be made and told her that was exactly what I intended. Somehow, thinking I would effectively be paid to put up with more animals made it seem quite acceptable. Emma wrapped her arms around me. Hugging her tight, I looked down at Roxi and hoped the baby father's genes packed a slightly smaller and prettier punch.

Creeping up on the rabbits.

My son getting down with the ex-batts.

Butch pulling a 'wee' face.

PART THREE

An eye for the camera.

Butch & Roxi, at home under the oak, summer 2010.

23

Tools for the Task

Each time I visited Tom at his smallholding, he made everything look so easy. Inevitably, I would measure my achievements against his. When I walked down to talk to him about breeding minipigs, I found him in the workshop. There, clasping a massive nail gun that looked capable of taking out alien invaders, he showed me his latest creation.

'It's a self-standing perch for my hens,' he said, as I admired what looked like a hurdle skilfully constructed from lengths of sanded wood. 'I can build you one, if you like. It won't take a minute.'

I didn't doubt his word. I'd have bodged something together from a pogo-stick, parcel tape and two box files. The job would've taken me the better part of a day, and lasted for less than an hour. Chances are Roxi would've gone on to eat the files. And Butch would've stashed the tape.

'Thanks all the same,' I said, declining his offer. 'But I'd really appreciate your help to electrify the fence. I need to be absolutely certain that the pigsty is secure.'

Tom rested the nail gun back against his shoulder. He still had his finger in the trigger. As I told him about our second overnight visit, it looked like he was set to hunt down the pair responsible and fix them to the nearest wooden panel. When I explained that Butch and Roxi had succeeded in disturbing them, he placed the gun on his workbench, much to my relief.

'Even if they wanted to get their hands on your minipigs,' he said, 'they'd face a battle trying to catch them. You've jumped on Butch one time. How difficult was that? As soon as anyone tried, they'd kick up enough noise to raise the alarm.'

Tom's outlook made sense, and I nodded throughout. Even so, I knew that I would feel better if we could bolster the defences. Emma was also in favour of an electric fence. I hadn't told her my real concern as to why I wanted to protect the minipigs. It seemed silly to worry her unnecessarily. With all her thoughts revolving around tiny minipiglets, her concern was simply for stopping a fox from paying a visit.

'There's one more reason why I need to take extra steps to secure the pigsty,' I admitted, glancing at the ground for a moment. 'If I'm lucky, it could be anything from eight reasons or more.'

Tom knew just what I was talking about. When I looked back up at him, I could tell by the way he appeared to be awaiting a punchline.

'Are you serious? You're going to breed that freak of na—' Tom stopped himself there, and found another way into his response. 'So, Roxi's going to produce some minipiglets, eh? She's mature enough, I suppose. But how did Emma persuade you into that one?'

'It was my idea,' I confessed. 'I made it with my business hat on.'

'Your business hat?' Tom leaned back against his workbench, arms folded now. 'Is it one with jingle bells attached?'

'That's a jester's hat,' I said, unimpressed. 'And this is serious.'

Keen to prove my point, I encouraged him to think about

the financial rewards. When I suggested some figures Tom almost seemed impressed. Finally, I pointed out that raising minipiglets from scratch would enchant my wife and children.

'My problem,' I told him, 'is that I don't know where to start.'

Tom dwelled on this for a second. 'Well, to begin you need a mummy minipig and a daddy minipig—'

'Who have to love each other very much,' I said to finish for him, though I was in no mood to play around here. 'I understand the basics, obviously. I also reckon I know how to tell when Roxi is brimming.'

'You should also look out for a swollen red vulva,' he said. 'It'll often become moist as well, which you can always check with one finger.'

I don't know if Tom really expected to engage me in conversation about the finer points of pig fertility. I just looked at him while wondering if he'd actually said that out loud. It all seemed so close to home, but in a clinical way. Nevertheless, it wasn't the female side of the equation that concerned me. As it turned out, apart from possible mounting difficulties, Butch would have been a perfect partner, given that they weren't related, as we had been led to believe; but as he'd been castrated that disqualified him.

'So, where do I begin?' I asked.

'Two options,' he said simply. 'You can "bring in the boar", as they say, and get her knocked up naturally, or you can do it yourself.'

Briefly, I wondered if I should just go and buy a book about the subject. Even if it wasn't as informative, it would at least be free from wisecracks at my expense.

'DIY is not an option for two good reasons,' I told him. 'As well you know.'

Tom looked confused. Then he seemed to pick up on why I had just been a bit short with him.

'I'm talking about artificial insemination,' he said. 'You'd be closely involved in the process, but not *that* close, obviously.'

It didn't take me long to weigh up the options. 'I think we'd prefer to get it done naturally. Bringing in a boar would be less messy, I imagine.'

The way Tom shook his head told me that was the wrong answer. 'In which case, you'd have to accommodate him for a month, to be sure he's present when she's brimming. It's also expected that you pay for his upkeep, with no guarantee that he'd do the business. On top of that you'd need to arrange to pick him up from the last customer, and deliver him to the next one.'

I listened to this and had just one question. 'What does the owner of the boar have to do in this exchange?'

'Collect the stud fee,' said Tom. 'It's just how it works in the world of pig keeping.'

'But that's money for nothing!'

'Isn't it?' Tom agreed. 'You've never seen my stud boar, have you?'

'I had no idea you even owned one!' I declared, stunned by this revelation. 'I've certainly never seen him.'

'That's because he's barely here. Tiger goes from one smallholding to the next, spreading his seed and lining my pocket.'

All of a sudden, I had another reason to bitterly regret having Butch castrated.

'OK, so I'm unwilling to pay for someone else's male minipig to wreck my garden for a month on the vague off chance that he'll impregnate Roxi. What's the deal with artificial insemination?'

'It's as easy as siphoning petrol.' Tom hopped off his stool and crouched to rummage through a shelf under the workbench. He returned with what looked like one of those long, bendy child's straws that finished with a spiral. 'What we have here is the catheter,' he told me. 'It's the main thing you need to deliver the goods. Mine came free with a bulk order of pig feed, but you can pick them up for a couple of quid at most.'

'Now you're talking,' I said, encouraged by such cost-saving measures.

Tom handed me the catheter to inspect.

'Naturally, you'll want to use a lubricant,' he added. 'It's also worth getting a good boar spray.'

I looked back at him. 'A what?'

'So you smell appealing.' Tom rubbed his wrists together and pretended to appreciate the perfume. 'It's a synthetic mixture that smells just the same as a substance produced by the boar to encourage the gilt's juices to flow.'

I crinkled my nose. 'So, it's a kind of eau de pig gland?' I said. 'That's not selling it to me, Tom.'

'Trust me,' he said. 'Apply some before you approach Roxi. It'll work like a charm. She'll be putty in your hands.'

I reflected on the wisdom of what I was getting myself into here. 'You make it sound like I need to seduce her first.'

'Do it right and she'll put out on the first date,' said Tom, in a way that made me think he meant it. 'Your task is to make sure the semen that goes in it is top quality stuff.'

'Naturally,' I said, returning the catheter to him. 'But does this mean I need to visit minipig breeders and ask to feel the boar's testicles? In those circles, is that kind of thing socially acceptable?'

Tom shrugged. 'There is an easier way.'

'Whatever it is,' I said. 'I'm interested.'

Tom reached for his jacket. It was hanging by a nail from the inside of the workshop door.

'Let's go for a drive.'

Twenty minutes outside our village lies a small trading estate. It's one of those places that barely caught my attention. Usually, whenever I drove by, I'd be too busy fiddling with the radio or yelling at the kids to notice it. As Tom swung his Land Rover onto the pitted asphalt forecourt, I noted that all the other vehicles here were either flat bed trucks or four wheel drives with attachments for livestock trailers.

'What is this place?' I asked, as Tom pulled up outside a pitch-roofed warehouse, in a space marked KEEP CLEAR. 'Are you allowed to stop here?'

Killing the engine, Tom smiled quietly to himself.

'It's my personal parking space,' he said. 'I'm such a regular here it would be rude to park anywhere else.'

I turned to look at the entrance. The solid-steel crash doors and industrial-sized trolley bay made it look like the most unwelcoming store I'd ever seen.

'Just a guess,' I said, 'but this doesn't look like somewhere that believes in advertising.'

Tom unplugged his keys from the ignition.

'You can't beat word of mouth,' he said. 'It's like a supermarket for smallholders. Members only, but you're good with me.'

As he spoke, the crash doors swung open. The imposing figure that had just used his trolley as a battering ram seemed unhappy to find his passage partially blocked by Tom's vehicle. He was sporting a full-length oilskin jacket, broad-brimmed hat and the kind of stubble you could use to strike a match.

I saw no sign of a rifle slung over his shoulder, but it would've been no surprise had he reached back for one and blown out the windows and tyres.

'We should find a proper parking space,' I said quickly.

'No need.' Tom waved cheerily. In response, the guy's scowl softened and he wheeled his way around the vehicle.

'Do you know him?' I asked.

'Of course not,' Tom replied, climbing from the driver's seat. 'But everyone around here knows *me*.'

I stayed close to my friend as we pushed through the doors. Inside, the store was arranged into aisles, with product banners hanging from the roof joists. Surprisingly, it looked like a supermarket in many ways. Where it differed was in the goods on display. Instead of tins of beans, fresh milk and vegetables, I found myself presented with offers for rifle racks, rabbit nets and stacks of monstrously big rat traps. I looked up and around, taking in displays of hobnail boots, industrial-sized bottles of bleach, rolls of chicken wire, claw hammers in six sizes and an array of restraining harnesses with huge clips and buckles.

'Is this really for smallholders?' I asked. 'It looks more like a place you'd come to stock up for the zombie apocalypse.'

Tom walked with a purpose, basket in hand, acknowledging everyone from customers to staff. If the people shopping seemed solitary and reclusive, those old-timers who were here to serve looked like they'd first had to seek permission to do so from their parole board. Several stocky, white-haired shelf-stackers wore their shirt-sleeves rolled to the elbow with faded tattoos on display. Like Tom, they looked weathered. Just not in a good way.

'Ex-poachers and gamekeepers,' whispered Tom as we walked behind them. 'When they're too old for the job, they find employment here.'

I glanced over my shoulder, unsure whether Tom was just fooling with me once more. On this occasion, I decided he was deadly serious. We walked on, passing a section for chainsaws, as well as sacks of animal feed stacked up at the rear of the store, and then switched back into the next aisle. When I saw the electrical fencing, box batteries and plastic stakes, I realised why we had come here.

'This is just what I need,' I declared, despite having no idea where to start.

To my relief, Tom picked off what was required. I had assumed we would then head to the checkout. Instead, he invited me to follow him across the store.

'There's more to come,' he called back, a little too loudly for my liking. 'If we're going to make progress on your spunk hunt.'

'*Tom*!' I shrunk into my shirt as one or two people looked around. 'He's talking about pigs,' I said to one, by way of explanation. '*Mini*pigs.'

'You should probably keep that bit to yourself,' Tom advised me, scanning the aisles from the front end. 'Progress isn't something that appeals to smallholders. They're traditionalists at heart.' He stopped there, and pointed at a section to our right. 'This is our next stop.'

The items on the racks weren't just designed to appeal to pig keepers. Anyone breeding livestock would find all the equipment they needed in order to do it for themselves. Tom picked off a bottle of sterilising solution. He tossed it into the basket, along with a catheter like the one he had shown me, and a canister that reminded me of a mouth spray. Without asking, I figured the latter item wasn't something intended to freshen the breath.

'Is that everything?' I asked.

'They don't sell lubricant. But I imagine you can source that for yourself.' Tom looked left and then right. 'There's just one more thing we need to check out.'

Before I could reply, he strode away at such a pace that I struggled to keep up. I really didn't want to lose him in this place. Every time I turned round there were people purchasing items of equipment that could somehow cause harm. From scythes to pitchforks and butchery blades, I wondered how many cult leaders were browsing at this moment in time, looking to protect their ranch houses from a raid by the authorities.

'Where are we going?' I asked, without meaning to sound quite so whiney. It reminded me of my young son, who would always be appeased on a supermarket run with a gingerbread man or cheese straw. Even if I was hungry, the only food products on sale that I could see were designed to paralyse vermin and liquidise their insides. As Tom led me into the aisle on the far side of the building, all I really needed was some fresh air. Until, that is, I saw what he wanted to show me.

Where a supermarket offered a section for popular lifestyle magazines, this place displayed a long rack packed with the kind of specialist titles that didn't need to be stocked on a top shelf. I scanned the magazines on offer; each one dedicated to a different aspect of smallholding, from growing your own to keeping chickens, doves, goats, ducks, horses, honey bees, llama, sheep . . . and pigs.

'Most breeders advertise,' he said, picking out a title with a huge pig on the front cover and the strapline, *Baconer Special.* 'It's just a question of looking through the small ads.'

I collected another copy, along with two other titles from the pig shelf.

'This is perfect!' I said. 'I never expected impregnating Roxi could be so easy. I can sort this from the kitchen table over a cup of tea and a biscuit.'

Tom glanced across at me from the pages of his magazine. He seemed a little pensive all of a sudden.

'This isn't something you should do with your eyes closed,' he said.

'Or the lights off,' I added, and then realised he wasn't joking. 'Sorry,' I said, and returned to inspecting the magazine.

For a moment, neither of us said anything. Tom picked up a title about shooting. I found myself looking at a problem page about pig keeping, but I was more concerned about my friend's change in tone.

'Are you one hundred percent sure you want to go ahead with this?' he asked eventually. 'I've no doubt there's a profit to be made from minipigs, but breeding can come at a price.'

'How so?' I asked, and turned the page.

'Mark my words. It's going to cost you time and test your patience.'

'How can you be sure?' I loaded my magazines into the basket. I felt confident in my decision, despite this moment. 'All your pigs come from the market. You've never bred them yourself.'

'For good reason,' said Tom, and closed the magazine in his hands.

I told him to add it to the pile for purchase, and led the way to the checkout. Having paid for everything, I even collected a form to fill in for membership and another for a loyalty card.

My decision to get into breeding minipigs paid dividends from the start.

As far as my children were concerned, it was the greatest thing I had ever done for them. Allowing them access to a laptop? Agreeing never to dance again at family weddings? Nothing came close to this. Best of all was the fact that Emma shared their excitement and gratitude. From the day I had suggested that Roxi could meet her maternal needs, my wife's spirits simply skyrocketed.

This was no empty promise, as they could see for themselves when I began to work my way through a stack of pig keeping magazines. Each title contained a comprehensive classifieds section for every region of the country. The sheer number of pig breeders in the south east of England was impressive. There were more than enough contact details for me to steadily decorate with red pen circles and teacup rings. Lots specialised in 'rare breeds', and I felt sure that minipigs would count among them. Peering over my shoulder every now and then, Emma took the view that I was basically spoiled for choice. What mattered, she advised me one morning as she left for work, was that I pinpoint a quality supplier.

My problem, I discovered, wasn't so much availability. It was the specific nature of my requirements . . .

'Hi. Yes. I'm calling about your advert in *Pig World Quarterly*. That's right. I'm looking to acquire some semen . . . you do? Oh, that's wonderful! Can I just clarify, before we go any further, this would be semen from a minipig . . . a mini— hello? *Hello*?'

This wasn't a one off. Every time I put in a call, the breeder would talk to me happily until I mentioned precisely what I had in mind. At best they would respond by sounding vague, evasive or suspicious. At worst, they would simply put the phone down on me. I couldn't understand it. My efforts left me feeling like I was seeking to acquire a slave child. What's more, as time pressed on, I found myself beginning to run out of contacts. I just didn't have the mental strength to start with another region, and so I placed all my hopes in those leads I had left. By the end of that week, having devoted my lunch hour each day to working the phone, I had just one circled entry to explore. The number wasn't the last on my list that I tried. It's just every time I'd called it, the line was either engaged or rang endlessly. The advert didn't offer me much. It read simply: *pig semen for sale. All breeds.* No contact name was provided. Just the digits for a mobile phone number. Until someone picked up, however, it was useless to me.

Naturally, I shared none of this with Emma and the kids. Admitting I had failed to get my hands on the goods was just out of the question.

'How are you getting on?' My wife asked me that weekend. 'I thought you'd have what you needed by now.'

We were talking in the front room. Once again, Emma was clutching a syringe. This time, she had no intention of worming herself. The scratch she'd gained during her encounter with Butch was only just beginning to calm down. It had been inflamed for much of the week, though Emma refused to see our doctor on account of the shame. In her view, it was nothing that couldn't be treated with an antiseptic cream and a brandy nightcap. Despite the fact that things hadn't quite gone to plan when it came to injecting Butch, she was keen to prove that she could do it properly. As soon as the replacement needle arrived in the post, she had volunteered to give Roxi her shot, which was what we were preparing to do when she asked for an update on the pregnancy project.

'A delivery should arrive any day now,' I said, hating myself on the inside. Aware that I wasn't good at lying, I diverted my gaze to the minipigs. 'Who's been eating biscuits in here?' I asked.

Butch and Roxi had been allowed inside again. Following Lou's sleepover we had granted them visitation rights under close supervision. In particular, it was something the little ones relished. This time, before they were allowed in to play with them, Roxi would require her shot. At that moment, both minipigs were rooting under the sofa cushions, sniffing out crumbs. Butch still had to climb up to do this, unlike Roxi who could stand with four trotters on the carpet and still reach the far corners. As well as leaving Butch behind in size, it was her bulk we had to watch out for. A cuddle on the sofa risked crushing injuries, and if she chose to flop onto the carpet it was a scramble to get our feet out of the way. Fortunately for us all, our grunting guests had begun to understand that the house was not a place in which to rampage. Butch appeared to have left behind his thieving sprees and Roxi hadn't chewed through a wire since the incident involving my gaming console. I had been grateful to Emma for sourcing a replacement wire so quickly, even though I knew she'd done it so the minipigs didn't get banished on a permanent basis. To her credit, despite fainting on the job last time around, my wife had also insisted that she would be the one to finish the worming procedure.

'It's like falling off a horse,' she said, on hovering over Roxi with the needle poised. 'You have to get back on it straight away if you're going to overcome the fear. The same goes for injecting pets.'

Without further word, she calmly inserted the needle in the space behind Roxi's ear and delivered the shot.

'You did it!' I beamed at her as she finished the job without once swaying. Roxi was so busy hunting crumbs that she even took it without complaint. 'You were brilliant. I couldn't have done that.'

Emma replaced the cap on the syringe. Despite her cool manner, I could tell she was elated. What's more, she didn't have a bad word to say about me.

'We're a team,' she said. 'I'll do the worming. You focus your efforts on making those little minipiglets.'

I really was digging myself into a hole. One becoming deeper by the day. Even Butch and Roxi couldn't match me in terms of how far down I was heading.

When I first started calling the numbers, I had simply assumed that my luck would turn. If only I'd been transparent about it with her from the outset, Emma would've understood. I couldn't be held accountable for the fact that nobody wanted to talk to me, after all. Instead, as I worked my way through the classifieds I had continued to assure my family that everything was going to plan. Was I close to getting my hands on quality minipig semen? Of course! It was just the post being slow, I told them. In return, I found myself living with excited kids who didn't squabble quite so badly and a wife who had last been this happy when compelled to wear maternity trousers and flat shoes.

I just couldn't figure out why my enquiries had prompted such a negative response. It was only the following afternoon, when Tom showed up to electrify the pigsty, that I had a chance to talk things through.

'You know your problem?' he offered, while screwing the fixings along the fence. 'I reckon this semen you're after is leaving a sour taste in the mouth of the breeders.'

I was standing behind him, handing out fixings as he needed them. 'Do you want to rephrase that?'

'Think about it,' Tom replied. 'These minipigs are worth a mint. Their stock value is soaring. Even I've seen pictures of them in the papers.'

'Let's not forget the celebrities,' I added. 'A celebrity without a minipig might as well abandon trying to further their career.'

'Which is forgetting how they're created.' Tom turned to face me. 'It's like parading around with the latest handbag that's actually been made in a sweatshop.'

'How so?' I asked.

Tom paused for a moment, like someone with bad news to break.

'The minipigs making headlines nowadays might look cute,' he said eventually, 'but they're basically the result of cross breeding runts through several generations. They're produced for size alone, rather than the quality of their meat. I've asked around. There's bad feeling about their popularity. It's the farming equivalent of a Frankenstein experiment, and this is the end result.' He gestured at Butch and Roxi. With the sun shining, the pair were baking happily under the eaves of the oak tree.

'They're not monsters!' I reasoned, feeling a little slighted. 'I know they can be testing at times, and grow a little bigger than most people realise, but at heart they're docile creatures.'

'No doubt,' said Tom. 'But rare breeds are a different pig entirely. We're talking Tamworths, Gloucestershire Old Spots, Mangalitzas; these are traditional, quality pigs that were once threatened with extinction in the face of intensive pig farming. Now, thankfully, they're thriving once more, but look at it from the traditional breeder's point of view. They're not in it to make a quick buck. They do it because they're passionate. For the sake of future generations, they're in the business of keeping the bloodlines as pure as they can.'

'So now these breeders aren't just weird with me on the phone,' I said. 'They're also Nazis?'

Tom looked less than amused. 'All I'm trying to say is that

when people like you call up to get your hands on minipig semen, they immediately mark you down as being another unscrupulous chancer hoping to cash in on the craze. Now, I know your intentions are honourable, but they don't know that. You're just trying to keep Emma happy, right?'

Awkwardly, I broke his gaze for a moment. 'The money had also crossed my mind,' I admitted.

Tom took another fence fixing from me. 'How many more contacts do you have left to try?' he asked.

I confessed that I had just one, a mobile number, and explained the difficulties I was having with it. 'When the line isn't engaged, it just rings and rings.'

'Sounds dodgy.' Tom began to screw in the fixing. 'Keep trying.'

I watched him at work for a moment, aware that we only had a few fixings to go before it was time to start wiring.

'Shall I put the kettle on?' I asked, despite knowing what his answer would be.

Inside, as the tea brewed, I found a note from Emma. She'd taken the little ones to the park. As I was with Tom outside, Lou and May had been allowed to stay behind. Stepping into the kitchen, it looked to me like the kind of place where all humans had been vapourised mid-breakfast.

'Girls,' I called up the stairs. 'These bowls won't wash themselves!' A damp towel lay in a heap at my feet. It had clearly been dumped over the banisters in the hope that somehow it would teleport itself into the dirty washing basket. 'You need to make more of an effort, OK?'

From upstairs, both girls chimed an apology, and promised to pull their weight. I picked up the towel, hesitated for a moment, and then dropped it back on the floor. Lou and May could deal with it, I told myself, and headed for the phone.

I had my own mess to sort out.

I didn't even have to punch in the digits. The 'recently dialled' section of the phone's menu was filled with the same number. Privately, I hoped to get an engaged tone. It meant I could ring off, having tried my best, and free my hands up to extract the tea bags from the mugs. I sighed to myself when it started ringing, and wedged the phone between my shoulder and ear. Previously, I'd let it ring for about a minute. The last time, in a state of stubbornness, anger and desperation, I'd gone to five. This time, it just seemed pointless. Nobody was going to pick up, and indeed that's just what happened. Instead, as I crossed the kitchen with two wet teabags on a fork, the answer machine kicked in.

By the time I handed Tom his cup of tea, he had already wired up half the fencing.

'What took you so long?' he asked. 'Not that I'm anything but fine working alone.'

I told him I had just enjoyed some level of success. 'The guy's got my number now,' I said. 'I'm sure he'll call back.'

Tom cradled his mug in one hand. He didn't bother using the handle, even though the tea was scalding hot.

'How did he sound?' he asked. 'Like a man who had some minipig semen to sell?'

'There was no welcoming message. 'It just beeped.' I reflected on this for a moment. 'That's dodgy, isn't it?'

He shrugged. 'Let's see whether or not he gets in touch.' Swigging his tea, into what must've been a gullet built to the same heat-resistant specification as his palms, Tom then returned to running the wire from one fixing to the next. 'As for your nocturnal visitors, if they do return it'll all kick off should they try to climb the fence.'

I watched him tie off the final fixing, before running the wire to the car battery I had bought. Tom had placed it inside the shed, before drilling a hole through the wall in order to complete the circuit. He attached the crocodile clips, switched on the battery and stood back. 'It's testing time,' he said to me. 'Go ahead. Try it out.'

'But I'll get a shock,' I said warily.

'Which will prove that it's working. The battery is on a low setting for now. It won't stop your heart. At least, let's hope not.'

I scanned the wire running around the top of the fence.

'It looks fine,' I said, and cleared my throat. 'You can tell it's on just by looking.'

Tom tutted to himself. Then, without a word, he grasped the wire. He didn't twitch or blink.

'You're right,' he confirmed, and let go again. 'It's on.'

'Is it?' I touched the wire with one finger, following his example, and experienced what felt like a mule kick through my solar plexus. 'Hell's *teeth*! That really hurts!'

Tom nodded to himself, seemingly satisfied by my response, and returned to the shed. 'I'll set it to maximum,' he said. 'If they do come back, they won't return again.'

'That's because we'll have killed them,' I stated, still reeling from the shock. 'You can't set it too high, Tom. It's just a deterrent. I don't want to see anyone murdered over a minipig.'

Tom looked a little disappointed. Reluctantly, he set the dial midway, and then closed the shed door. I was just helping him to pack away his tools when Lou sauntered out to join us.

'Hi,' I said brightly, as she stopped behind the gate, arms folded and with a look of bored disdain on her face. 'Have you tidied up yet?'

My eldest daughter shrugged. 'I'm waiting for May to help me,' she said. 'Only Miso has fallen asleep on her lap and she won't move until he wakes up.'

Tom and I exchanged a glance.

'I'd invite you to help us,' I said. 'But we've just finished. Lou, you have to promise me you won't touch the fence. It's live now.'

Lou nodded, but her attention had already drifted to Butch and Roxi. Maybe that's what jogged her memory, because all of a sudden she revealed her reason for venturing outside.

'A man just called,' she said. 'Something about minipigs? I told him you were busy.'

'*What*?' I took a breath. 'Lou, that call was urgent. You should've fetched me.'

Lou looked at me as if I had just questioned the reason for her existence.

'I was only trying to make myself useful!' she said. 'Can't I do anything right?'

'What else did he have to say?' I asked.

'Just that he would call back. He wouldn't leave a name or anything. To be honest, Dad, he sounded shifty.'

Despite her concern, Tom beamed and clapped me on the back.

'Sounds like you've found your man,' he said.

24

A Man about a Minipig

In terms of tricks, Butch and Roxi were hardly candidates for the circus. Emma may have persuaded me that these were creatures who could practically sit down over a chess-board. In truth, they would just eat the pieces. The minipigs were smart. They just weren't prepared to use their brains and ingenuity unless food was involved. I'd already had to remove the metal bin from the pigsty. It contained a sack of pig nuts. Both had come from the smallholder's supermarket. I had become a regular customer, with a loyalty card that was building up points rather nicely. The food was cheaper than the starter bag sold to us by the breeder, and the minipigs seemed to prefer it. In fact, this was Butch and Roxi's staple diet. They had one meal at the crack of dawn, if only to shut them up, and then another one mid-afternoon, after which they'd wind down for the day. Until, that is, they started tipping over the bin and nosing off the lid.

Then they moved onto the bird feeder.

In order to pretty up what basically looked like a section of the Somme, Emma had hung one from an oak branch over-hanging the pigsty. This proved a popular attraction, and not just for the squirrels. Very quickly, the minipigs worked out that activity on the feeder would lead to seeds dropping to the floor. As soon as a squirrel fancied helping itself to a snack,

Butch and Roxi would race for position directly underneath. After some practice, it meant those seeds that fell from the feeder wouldn't actually reach the ground.

It was only after Tom had secured the pigsty that the minipigs showed their true ingenuity. With a current circling the perimeter fencing, I felt confident that anyone who attempted to gain entry would be punished with one hell of a surprise.

What I hadn't anticipated was the shock that *I* would experience when the minipigs learned to lift the gate latch and let themselves out.

The first I knew of it was when the cat flap crashed open. I was in the front room at the time, having forced the little ones to switch off the television and make a puzzle with me. The force with which the Perspex disk smacked upright against the inside of the door was enough to startle us all. Seeing Roxi's pink snout push through was what caused me to gasp and rise to my feet. With Butch behind her, it looked like they were queuing to be let in.

The first two times it happened, I marched them both back before they'd had time to destroy the garden. This was easily done. So long as I got out quickly, and scattered some pig nuts across the concrete, Butch and Roxi would return at a brisk pace. On Roxi's third appearance at the cat flap, I realised something would have to be done about the latch. At the same time, I was tired of shepherding them back to the pigsty, and so I headed for the back door in a bid to try something different. Collecting two pears from the treat tin outside, I returned to the front room where Roxi's snout was still visible through the cat flap.

'Take it,' I said, waving one pear in front of her nostrils. 'It hasn't been in the kitchen. You're good to go.'

When the snout withdrew, I followed after it with the pear until it was relieved from my grasp. This, I repeated for Butch, and then watched from the glass as both he and Roxi trotted back to their pigsty. They carried their prizes inside their sleeping quarters, and disappeared from sight. Then, knowing that closing the gate was basically child's play, I persuaded the little ones to do it for me. In return, I offered to finish the puzzle quickly so they could go back to watching the television.

Fixing the gate was simple. I just barricaded it using the feed bin. I did intend to head down and see if I could do something that wasn't basically a bodge job. But as the working week began, I didn't want to risk missing that all-important call. I regretted only leaving our home number with the breeder. It meant every school run became a mad scramble to get back. I even took the phone off the hook each time I left. I figured if he called and got the engaged tone, rather than an answer machine, he would be more likely to call again. Several days passed in this way. Every evening, Emma would ask me for news, and I would make positive noises, by which I mean that I lied to her outright.

'The goods were sent by second class post,' I actually assured her one time, which filled me with self-loathing. 'Any day now, we'll be in business.'

When it didn't show up, I did at least have an excuse to hand. As I wasn't entirely sure how legal it was to be introducing pig semen into the postal system, I felt it was reasonable to tell Emma it had been confiscated. All the time, I hoped and prayed the breeder who spoke to Lou while I was outside would stay true to his word and ring back. I had the man in mind one morning in my office. I should've been working at my keyboard. Instead, I was gazing out of the window; fretting inwardly. A knock at the front door drew me from my thoughts. I even grasped at the hope that by some miracle the breeder had known the purpose of my call, worked out where I lived, and was here to hand deliver the goods. Instead, on opening the door, I found myself facing my neighbour.

'Roddie,' I forced a smile. 'Good to see you. If it's about the electric fencing, I know it might look bad from your side, but I see it as another line of security for us both.'

My neighbour was clutching a sheaf of home-produced pamphlets in his arms. I recognised them straight away. We lived in one of those villages that still produced a black and white parish newsletter every season. Roddie handed one to me.

'I recommend you read this closely,' he said. 'Something must be done.'

I opened the newsletter, which was adorned with a hand-drawn portrait of the village church. I couldn't read the artist's signature. Judging by the talent on display, however, I figured it had been a recent transition from crayons. From experience of previous issues, I wasn't looking for details of a drive-by shooting, a drugs bust or the discovery of a hell mouth midway between the pub and the primary school. In this village, the very definition of 'news' was open to interpretation.

'What have I missed?' I asked. 'Have the flowers outside the hall been trampled on again?'

'It's serious,' he replied. '*Thieves* have been spotted in the area.'

The way he said this told me just what kind of threat Roddie had conjured in his mind. Were he to complete a photo-fit, I anticipated we would be on the look out for men in stripy tops with swag bags over their shoulders or snaggle-toothed gypsies in horse-drawn caravans. My neighbour lived in fear of stereotypes, but then it wasn't my place to judge him. I was also aware that I had recently witnessed suspicious activity with my own eyes. Not that I had any intention of stoking Roddie's fears by telling him about the two figures in the lane at night. They were only interested in the minipigs, after all, and I had taken steps to punish anyone who dared to steal them, vigilante-style. Instead, I thought I'd take advantage of the situation by winning him over in advance of any future antisocial behaviour from Butch and Roxi.

'I'm sure there's no need for alarm,' I assured him, and offered what I hoped would come across as a winning smile. 'If you see or hear anything suspicious, you can always turn to me. I'm here for you, Roddie. I really am. Even if you just want a chat. It's what good neighbours are for.'

As I said this, the house phone began to ring.

Roddie seemed not to register it. He looked restless and uncomfortable on my doorstep, still troubled by the robbers prowling through his imagination. It was clear he wanted to talk. I knew the decent thing to do was offer him my listening ear. I just desperately needed to press the phone to it first.

'I don't mean to be a burden to you,' he said.

'We would never think of you like that,' I heard myself reply, despite being focused entirely on the sound coming from the kitchen. I began to retreat from the front door. 'Won't you come in?' I suggested, aware that if I didn't pick up within the next few seconds I could lose the caller to the answer machine.

'You're very kind,' Roddie replied, after what felt like an age. 'But I really must continue with my deliveries.' My neighbour gestured at the newsletters he had yet to deliver. 'It wouldn't do to find myself at the far end of the lane towards sunset.'

It was the break I needed. At any other time I would've reasoned that there was a difference between a thief and a vampire. As the worst crime I had witnessed in this village concerned the dropping of a cigarette packet from a passing car, I really didn't think I was allowing him to head out into the badlands here. By now, the phone must've been just one ring away from the answer machine. As Roddie turned to leave, I was at the door bidding him goodbye within a nanosecond. By the time I reached the kitchen, the tail end of my welcome message had just played out. I practically dived for the phone, hitting the pick-up button at the same time.

'Don't go!' I cried. 'I'm here!' At first, it seemed that I was too late. I heard nothing from the other end of the line, just a crackle of static, to begin, and then, to my delight, the sound of someone breathing. 'Is that the breeder?' I asked, and beamed to myself when the caller cleared his throat to confirm.

Our conversation was brief; so swift that I only really digested the nature of the deal after I'd set down the phone. Finally, I had located what, to my mind, was the elixir of pig keeping. I just couldn't share my news with Emma. As far as she was concerned, I'd already arranged to collect from my fictional source. The breeder wouldn't give his name, but that was fine by me. He had what I needed, sounded knowledgeable and even positive about minipigs. When I suggested a pick-up, he had given me the address for a pub in a rundown seaside town on the south coast, some forty miles from here. He insisted that we meet there, before moving on to his farm.

Then he had named his price. In cash only.

It was more than enough for me to hesitate. For a moment,

I had considered backing out. I only needed to remind myself that any more excuses wouldn't wash with Emma. I needed to find my voice and agree to the figure.

'So,' said Tom, as we pulled up opposite the pub in question. 'Are you going to tell me how much this stuff is costing you?'

I unplugged my seat belt and faced him. It was a detail I had avoided sharing when I first asked my friend if he would accompany me. This time, it wasn't so much because he knew all about pigs. Following the conversation I'd had over the phone, I was a little bit worried about my safety.

'Let's just say it all boils down to supply and demand,' I said. 'Once Roxi has her minipiglets, I'll make back the outlay and earn a profit in the process.'

Tom turned his attention to the pub across the road. With paint peeling from the frames, and one window boarded up, it didn't look like the kind of place you'd stroll into and order a nice glass of chardonnay. It was located just off from the sea front, where gulls haggled for discarded chips. There weren't many people around. By and large those vehicles on the road were mobility scooters. The only individual not huddled under a bus shelter or carrying thin plastic bags, was standing at the shoreline in a suit. He looked like he might be about to leave his clothes on the pebbles and wade into a new life. It was, without doubt, a grim place to be.

'This breeder,' said Tom. 'Do you know much about him? What's his form with minipigs?'

I shrugged in my seat. 'No idea. All he really said was that he would recognise me.'

'What? So, he's a fan of your children's books?'

I smiled to myself and gestured at the pub. 'That place looks like it's home to regulars only, if you know what I mean. I reckon as soon as I step through those doors the place will fall silent.'

Tom jabbed a thumb over his shoulder. 'There's a crow bar in the back somewhere, if you think it'll come in useful.'

I chuckled and showed him my palms. Then I faced him, and realised he wasn't joking.

'We're not Mafia,' I reminded him. 'Come on. Let's get this done.'

I let myself out of the Land Rover, inhaled the sea air and zipped up my jacket. I had expected to hear the driver's door opening, too. When that didn't happen I turned and peered through the passenger window.

Tom was still in his seat. I watched him switch on the radio and then settle down with the back of his head cushioned in his hands. I rapped on the glass. He glanced across, seemingly surprised to see me again already, and rolled down the window.

'What are you doing?' I asked.

'Waiting for the weather forecast.' Tom looked mystified by my interest. 'If that's OK with you?'

'I thought you were coming with me.' I gestured at the pub. 'Look at the place! I need you now more than ever!'

Tom glanced at the building across the road and then back at me. I waited for him to say something, but first a knowing smile eased across his face. As it did so, I realised I was on my own.

'We've come a long way,' he said. 'But this is your moment.'

The last time I walked with this much cash in my jacket pocket, I was buying a second hand car.

Even with Tom watching me, I felt vulnerable as I crossed the road for the pub. I had at least made him agree that if I wasn't out in ten minutes then he should come and find me. Tom had insisted that when it came to the purchase I had to trust my instincts. In a way, I took his point. If my friend oversaw the deal, and Roxi went on to give birth to mutant pigs, I would probably take the view that he should've known we were handling bogus semen. I had to take responsibility for my actions, here. I still felt way out of my comfort zone though, which was usually restricted to a six-foot semi circle around the telly at home. In a bid to stop myself from bottling it, I reminded myself that I was doing this for the sake of my family's happiness.

At the door to the pub, I looked at my shoes, muttered words of encouragement, and made my entrance.

Inside, it wasn't as bad as I had feared. The hum of conversation didn't fall to the floor when I entered. Nor did anyone look at me murderously or quietly open their flick knives. This was because there was barely anyone in the pub. A couple of old-timers were leaning against the bar. At a table in the corner, a bullet-headed man with no neck and a full pint was reading a tabloid newspaper. A clapped-out old dog sat at his feet, seemingly guarding a cooler box. The dog was the only one in the pub to lift his eyes in my direction. It looked utterly depressed, as if this was how it spent its days. For a moment, I wondered whether its owner was the man I had come to see. He was a big fellow, practically bursting out of a shiny red football top, with thick stubble shadowing his chops. I figured the cooler could only contain a furtive supply of beer. My pig-breeder radar didn't even twitch. Unwilling to look like I was scoping everyone out, I headed for the bar and ordered a sparkling water. It was lunchtime, after all. The landlord, another big man in a capped white shirt and a drying up cloth over his shoulder, looked a little surprised.

'Do you want anything with that?' he asked.

I glanced sideways. The old-timers were drinking pints of ale. I hated ale.

'And one of those,' I said, which seemed to satisfy him.

I looked around, aware that I could hear other voices. It was then I realised that a small group of women were at a table by the front window, hidden behind the door. Unlike the others, they were watching me with interest. My first impression was that these were office colleagues on a lunch break. One of the women, in her twenties with long, pinned hair, had a laminated menu in her hands. She whispered something to the others just then, which prompted much giggling. With my pint served, I raised it politely, toasting their high spirits, and took the slightest of sips. Looking around again, I discounted the men at the bar as I had Mr Soccer Shirt with the desolate dog. Maybe my contact was late? I checked my watch. On the phone, he had been adamant that I showed up on time. I sighed to myself, before setting down my glass in favour of the water.

'So, that's bacon sandwiches all round, right?' The voice came from the table behind the door. I faced around once more to see the woman with the menu had just stood up. She made her way across to the bar, and stood right beside me. I gave her some room. She smiled, without meeting my gaze this time, and yet there was something about her manner that told me she wasn't just here to order lunch. Was the thing with the bacon sandwiches some kind of code? I had expected to meet a man, but that was never spelled out to me on the phone. I was told, simply, that I would be approached. Hesitantly, I cleared my throat.

'How are the bacon sandwiches here?' I asked, as the barman finished serving one of the old men.

She looked across at me. 'Good,' she said. 'A little on the small side, but you've come to the right place.'

That was more than enough for me. I waited for the landlord to turn to the till. Then I leaned in beside her, anxious not to be overheard.

'I'm here for the semen,' I said quietly, though it was enough to command her full attention. 'The cash is in my pocket.'

The woman blinked in response. When she took a step away, backing into the old-timer behind her, I just knew I'd approached the wrong person. This wasn't a minipig breeder, or even a representative. This was a young lady ordering lunch. By now, the landlord had turned to serve her. He seemed surprised to find her looking at me as if I'd just flashed her demon eyes.

'Everything alright?' he asked.

'So sorry,' I said, not just to the woman but the landlord and the old-timers, too. 'I thought you were someone else.'

'*Who*?' she replied, sounding shocked to the core.

At once, the atmosphere seemed to sour. I glanced around, lost for words. I was hardly going to explain the true purpose of my presence to these people. Instead, in a very small voice, I said, 'I'm here to see a man about a minipig.'

I figured it would be several minutes before I could rely on Tom to extract me. Feeling as if the walls were beginning to close in, I turned to make a hasty exit. As I did so, the man in the soccer top looked up from his paper.

'Then you'll be looking for me,' he said, and gestured at the chair opposite.

Mumbling my apologies, I ducked from the bar and shrunk into the seat. The dog at his feet offered me a longing look, as if perhaps it thought I was here to take it outside and shoot it. I glowered at its owner, despite his imposing bulk.

'Why didn't you say anything when I first walked in?' I hissed, feeling annoyed as well as relieved.

The man gestured at his tabloid. It was open at the sports section.

'I was catching up on the scores,' he said, before folding the paper away. 'So, you want to make some minipigs?'

'I had no idea it would be so difficult to start out,' I told him. 'How long have you been in the business?'

The man shrugged, which I think was supposed to be an answer, before glancing around to be sure we were no longer being watched.

'I can let you inspect the merchandise,' he said.

Furtively, he removed the lid from the cooler. Reaching in, he produced what looked like three drinking straws.

'Is that it?' I asked. 'I was expecting a bottle or something.'

'It's frozen,' he told me. 'I can deliver fresh, but it's only good for a few days. By keeping it on ice, you can use it at your leisure.'

'So what's with the straws? Do I have to suck it out like snake venom or something?'

The man masked a smile, though I had been deadly serious. If I was right, it would be a deal breaker for me. I wasn't *that* committed to the project.

'What I have here are three hits,' he continued. 'That's three chances for you to hit the spot. When your minipig is good to go, you thaw one out and off you go. It's quality gear,' he said, furtively fanning them in one hand. 'I can guarantee it.'

I looked at the straws, feeling totally out of my depth.

'Don't take this the wrong way,' I said, 'but how do I know that what you're selling me is the real deal?'

The man seemed taken aback. 'This is weapons-grade, my

friend. Especially prepared for maximum impact. I give you my word.'

I was listening to what he had to show and tell me. It's just I was processing it with a slight delay. 'This is minipig semen, right? Not plutonium.'

Now he seemed disappointed, hurt even, and withdrew the straws from view.

'If you're not serious,' he said, 'I have other clients.'

'Wait!' I leaned in across the table; anxious not to cause a scene. 'I mean business. Really, I do. It's just you have to admit this all feels a bit . . . unconventional.'

If I had caused offence by suggesting the straws might not contain the genuine article, my candour about the circumstances of the deal had the opposite effect. At first the man just glared at me. Then he nodded to himself, ever so slightly, but enough to make his chins contract and expand.

'Fair enough,' he said, and produced a mobile phone. 'It's a lot of money. I understand that. I just need to protect my reputation. If other breeders discovered I'd branched out into minipigs, they could stop hiring out my rare-breed boars in protest.' The phone had a touch screen. He tapped it as he spoke, searching through the menu. 'Personally, I don't have a problem breeding hybrids. So long as they're healthy, and I can look the buyer in the eye and know he has their best interests at heart.'

Finally, the man found what he had been looking for. As he turned his phone to me, I took a very different view of who I was dealing with here. Having come across as the kind of person who seemed more likely to sell pirate DVDs, it really did appear that he had some integrity. Even before I looked at the video clip he had lined up for me, I felt better about doing business with him.

'Wow!' I said, as I viewed what he had to show me. 'Those are cute minipigs.'

'My boys,' he said, with what sounded like a hint of pride. 'They might be pint-sized, compared to my rare breeds, but they never fail to knock up the little ladies.'

He laid out the straws on the table, having checked once more that we weren't being observed, and then looked at me expectantly. I didn't like to think about how he had extracted the contents. What mattered was the fact that his livestock appeared to be the genuine article. I felt sure of this because the man's gloomy dog could also be seen in the clip. While the focus was on the three boars, rooting happily in straw on the floor of a barn, the dog could be seen moping about in the background. It looked like its life had been overtaken by minipigs, and here I was simply feeding that demand. Even so, judging by the appearance of this trio, I would be breeding from a fine stock. When the clip came to an end, the man pocketed his phone but left the straws on the table. I looked up, and found him waiting for a response. I patted the money in my pocket, just to check it was still there.

'I don't know,' I said, hoping this would prompt some kind of discount. 'I'm still taking quite a risk here.'

The man narrowed his gaze. 'If you're asking to see the minipigs for yourself,' he growled, 'it can be arranged. My truck is parked nearby.'

That wasn't what I had in mind. Having seen the clips I figured the minipig semen was legitimate. I was just hoping that by being awkward he'd be prepared to negotiate. Now I found myself faced with being obliged to travel alone with him. In a way, the risk to my personal safety seemed even more immediate than any chance that I was being sold a dud. The man was an imposing presence. If he wanted to imprison me, make me his plaything or just feast on my liver and spleen, I would stand no chance. He was waiting for me to answer, but the longer I hesitated the more insulting it seemed. Feeling like I had no choice, I drew breath to invite him to show me the way. As I did so, his eyes switched to the entrance behind me.

'Tom!' he said, sounding surprised but cheery. 'Alright, mate? What are you doing here?'

I turned to the figure at the door. Tom looked equally taken aback. He also seemed to relax visibly when he realised this was the man I had come here to see.

'Hello, Barry,' he said without moving. 'It's been ages since I've seen you at market. I didn't realise you'd moved into minipigs.'

The man called Barry seemed suddenly pained at the very mention of the word. He looked from one side to the other, moved to remove the straws from view, and blushed horribly.

'Tom's a friend,' I said, as Barry looked increasingly busted. I had certainly heard enough now. Tom had seen to that. If he could vouch for Barry, even if the man was compromising his reputation here, that was good enough for me. All that was left for me to do was make him an offer that would buy both the semen and our silence. 'If you're prepared to drop the price by ten percent,' I said, 'we have a deal.'

Tom approached our table. Barry glanced up at him. The straws were in his hand now, but even he could see that there was no point in hiding them. Tom nodded like it was a deal he couldn't refuse. Barry contemplated the straws, working things through.

'This didn't come from me,' he said finally, and offered his hand to seal the deal. 'If anyone asks, I breed pedigree pigs, alright? Just pedigrees.'

From the floor beside his chair, the dog let out a long, defeated-sounding sigh.

25

Short Straws

I was in business. With everything needed to impregnate a female minipig, all I had to do was wait for Roxi to start brimming.

Naturally, I anticipated some hitches. At feeding time, first thing each morning, I planned to push down upon her rear haunches, just as Tom had advised me. Even then, I wasn't entirely sure that I could identify the three-day window in which she was brimming. I really didn't want to start paying special attention to her vulva. For one thing, a lot of horse riders used our lane, with a clear view over the fence. As for the procedure itself, assuming I had worked out when Roxi was fertile, I knew it wouldn't be as simple as Tom had implied. Looking at the catheter, and the temperamental beast into which it was supposed to be inserted, I realised it would require some practice.

What I didn't foresee was the fact that Emma would be responsible for creating the first hurdle.

'What's this?' she asked, the evening after I had returned from the coast.

We were in the kitchen at the time, working out what to cook for supper. I turned from the cupboard to find her facing the fridge.

'Is it the leftovers from my lunch?' I asked. 'The dog can have that.'

Emma turned to face me. Looking at what she held in her hand, I realised she was talking about something she'd found in the freezer compartment.

'I think they're lollipops of some sort,' she said, holding one straw vertically for inspection. 'Did you make them with the kids?'

I shifted my weight from one foot to the other. 'Actually, that'll be minipig semen.'

For a moment, Emma looked so surprised I thought she might fling the straws to the floor. Her mouth formed a word, but she didn't make a sound. Then again, she might've just been gagging.

'So, this is it?' When she did find her voice, it was raised by several octaves. 'Why didn't you tell me it had arrived?'

'It was meant to be a surprise,' I reasoned, aware that it had been exactly that, but not in the way I intended. 'A *nice* surprise.'

'But did it have to go in the freezer?' Hurriedly, she offered me the straws. 'Here. Take them.'

'It's only a minor inconvenience.' I hoped that a smile would soften her outlook. 'Think of it as minipiglets on ice.'

Emma looked repulsed. 'Take them *away*! I am as keen as you for Roxi to conceive, but this . . . this is unreasonable behaviour! Any judge would back me, Matt. I mean it!'

I found her response a little extreme. It felt like I had been to the ends of the earth and back for these straws. As far as I was concerned, all three needed my undivided care and attention.

'But we have to keep them frozen,' I reasoned, keeping my voice down in a bid to sound measured and informed. 'If they thaw out too soon, the contents will . . . I don't know, curdle or something.'

Emma grimaced to herself. 'Is that what happens if you leave it out? The stuff *curdles*?'

'I'm no expert,' I said with a sigh. 'We should treat it like chicken breasts, I think. You know? Defrost thoroughly before use, but don't leave it at room temperature for too long.'

Emma closed the fridge door. 'I'm not hungry any more,' she grumbled, and left me for the front room. 'Do we have any wine?'

This semen was not for spoiling. After everything I had been through, I wasn't prepared to let it thaw out prematurely. Fortunately, I knew that Tom kept a little fridge in his workshop. An emergency call to his mobile was quickly followed by a dash down the lane to his smallholding.

'Steady on,' he said, as I crashed into his workshop. 'It's not like you're transporting pig valves for a heart transplant.'

At his bench, Tom was drawing up plans for something complicated. It looked like a blueprint for a bank heist, with crosses and numbers marking out exit points and security cameras. Knowing my friend, it was more likely to be a proposal for a water garden or a sundeck. Lately, he had been working into the evenings. The man was in demand; mostly for garden redesigns, but then here I was with an urgent need for appropriate cold storage. I had left the pub with the straws in the cooler, sold to me as part of the package. Now I'd been forced to stash them in there again, along with a bunch of ice blocks. I placed the cooler in front of me, and looked at Tom pleadingly.

'This stuff might not be a vital organ,' I said, 'but it certainly came from one, in view of the price I paid.'

Returning to his sketch, Tom invited me to use the little

freezer compartment. The fridge itself wasn't very big. It looked more like a hotel minibar. I expected to find it stacked with fish bait and beer cans. What I saw instead made me gasp and shrink from the shelves.

'Oh, that's disgusting.'

Finally, Tom set the pencil down and climbed from his work stool. As he turned to face me, rising to his full height, I stepped back on instinct. All the shelves had been removed to accommodate just one item. The dead pheasant sported a beautiful, coppery plumage that was only marginally overshadowed by the fact that it had been comprehensively flattened under a vehicle tyre.

'Roadkill,' he said casually, and then winked at me. 'My kind of meals on wheels.'

'You can't eat that,' I declared. 'It's been squashed.'

Tom shrugged. 'Once it's plucked and gutted, I prefer to think of it as being pan-ready,' he said. 'All kind of wildlife gets run over on this lane. Sometimes a deer if I'm lucky, though that requires a lot of sawing. Don't tell me you've never taken advantage?'

I reminded him about Misty the cat. When the vet had presented her to me, wrapped up in a towel, my first thoughts had not involved the stew pot.

'In a way,' I added, 'her death is indirectly responsible for the fact that I'm here right now, asking for some space in your freezer.'

'Don't let me stop you.' Tom gestured at the box still containing the three straws. 'Whenever you want them back, just help yourself. The key is under a brick behind the water trough.'

'Is that safe? Shouldn't you have better security measures than that?'

'I would lay man traps, and then lie in wait with a chain saw, but I'm not sure that would work for me as a case for self-defence.'

'I was thinking a lock on the doors would be a start.'

Tom shrugged, evidently unconcerned. 'It's not a problem. All my tools are engraved with my name. If anyone ever helped themselves I could always track them down.'

I wasn't sure if he was talking about his tools or anyone foolish enough to steal them. With more immediate matters on my mind, I returned my attention to the fridge. This time, trying hard not to glance at the pheasant, I slotted one straw after another into the freezer compartment.

'Thanks,' I said, on closing the fridge door. 'Not just for this, but for coming with me to the coast. I can't tell you what it means to me now I've actually got hold of this stuff. Without Barry, my life wouldn't have been worth living.'

'Then let's hope his boys don't let you down.' Tom reached for an eraser, evidently too busy to stop working.

There was something odd about the way he said this. It was as if he thought I had bought myself into something that was going to disappoint.

'Everything is above board,' I reminded him. 'We swapped paperwork and everything.'

'That's OK, then.'

Still Tom seemed like he was holding something back from me. 'You think all this is a bad idea, don't you?'

'No.'

Tom continued to sketch. I really wanted him to explain himself.

'You don't have to pretend,' I said. 'I know you've been a big help, but I can tell you have doubts. Level with me, Tom. What's the problem?'

Finally, Tom set his pencil down and climbed from his work stool. As he turned to face me, I stepped back.

'I bet Emma and the kids can't wait until the minipiglets arrive, right?'

'Absolutely,' I said, still puzzled. 'As far as I'm concerned, we all make something from this venture.'

'Which is my concern,' he cut in. 'What if Roxi doesn't get pregnant?'

'That's not an option,' I told him. 'After everything I've done to get this far.'

Tom sighed to himself. 'With a bit of practice, artificial insemination isn't difficult to perform. It's what happens afterwards that's out of your control.' Tom paused for a moment. 'All I'm asking you to think about is how will your family react if things don't work out? It seems to me you've got a lot more riding on it than you realise.'

All of a sudden, it felt as if Tom was interfering in something really quite personal to me. I sensed the muscles in my jaw line tightening as he spoke. Part of the problem was that I couldn't argue with his point. I had sold my family a dream: a house teeming with tiny minipigs. Not only that, it was a litter that would earn us the kind of holiday currently beyond our reach. If it proved too good to be true, I would be responsible for causing nothing but disappointment.

'I had better get home.' I retreated towards the door as if the atmosphere between us was forcing me out. As I did so, I picked up the empty cooler box. On rising once more, I met Tom's gaze one final time. 'You're wrong about that semen,' I added, feeling stung and grasping for the last word. 'I know quality when I see it.'

* * *

That night in bed, listening out for intruders, I dwelled upon what I should do for the best. It seemed like a smart move to encourage my wife to prepare for the possibility that mini-piglets might not materialise. It also felt like the surest way to burst the bubble of happiness that had enveloped her since I first suggested that Roxi could become a mother. I just couldn't bring myself to do it. The next morning, as Emma got ready for work, I drew breath several times to address the issue. On each occasion, it transformed into a troubled sigh.

'Everything alright?' she asked eventually.

Emma was just climbing into her coat at the time. The little ones were clamouring for a hug goodbye. It was all part of the morning routine, something Emma relished before setting off for work. As soon as she crouched, playfully screwed her eyes shut and spread her arms, Honey and Frank leapt into her embrace. When she finally looked at me, my mind was made up.

'Everything is fine,' I said, and dipped down to kiss her goodbye.

'Be sure to pass that on to Roxi,' she said. 'For good luck!'

Since announcing the plan to breed Roxi, my family had returned to looking at internet images of minipigs frolicking in flowers. Things had changed since these pictures last popped up on the screen. Not only had they exploded in number, to my eye they bore little resemblance to the hefty creatures that had destroyed the bottom of my garden.

The next morning, taking a break from work, I ventured out to the pigsty with Sesi at my side. If Butch had grown a little bit, Roxi was about as big as my dog. There was no way now that she could trot freely underneath her. Technically speaking, this was one mighty big minipig, but it was too late to return her for a smaller model. Even Sesi had become reluctant to join me in their space. To be fair, it wasn't just Roxi who made her wary. In return for protection from foxes, the chickens had taken it upon themselves to make life as comfortable as possible for their two bedmates. As well as providing them with eggs, our hens were quick to check out anyone who came through the gate, and that included the dog. Understandably fed up with having her paws pecked, she chose to settle on what was left of the grass while I let myself in. In the absence of a body scanner, the chickens hurried over to look me up and down with their beady eyes. Once I'd passed security, I was clear to meet the minipigs.

'What have you found now?' I left the hardstanding to pick my way across the mounds and troughs of earth. The crater they had excavated was so big that I could barely see Butch. Roxi was standing with her back to me, tugging at another tree root. This one wasn't spindly, like so many they had snapped. It was huge. Warily, I glanced up at the overhanging branches of the oak. 'I'm not sure this is such a good idea,' I said. 'What you need is something to distract you. Like minipiglets.'

I had come here to undertake what would become, over the weeks that followed, a daily and dispiriting check. This first time, Roxi made it easy for me. With her rump in the air, swaying from side to side, all I had to do was place my hands upon her flanks and push down. She didn't like it one bit, and rotated herself clear.

On those occasions when Roxi didn't have her snout buried in the ever-growing hole, I would throw some pig nuts on the ground before assessing her fertility. This, I learned to do having once attempted to manoeuvre myself behind her when she wasn't distracted. Instead of complying, she had moved backwards in surprise and pinned me to the fence. With food on the floor, I could carry out the check without a problem. Every time, however, Roxi showed no sign whatsoever that she was brimming. Every now and then I wondered whether I was doing it correctly. I knew that Tom would be able to advise me. I just felt uncomfortable about involving him in something he believed could bring me nothing but trouble.

After almost a fortnight of daily pushing and shuffling, and in a bid to understand the inner workings of our minipig, I turned to Emma instead.

'Roxi should be brimming for three days every three weeks, right?'

We were in the bath at the time, facing each other over a mountain of bubbles.

'That's such an unnecessary word,' said Emma. '*Brimming.*'

The little ones had been in bed for an hour. In that time, it had been twenty minutes since Frank or Honey last popped out, which was usually a good indication that they had nodded off. Downstairs, Lou was finishing her homework at the kitchen table. May was in the front room. She was watching television with the sound down low in case it unsettled the cat on her lap. As Emma liked to soak away her working day, I often climbed in with her if we wanted to talk uninterrupted. She liked the temperature of the water to be somewhere close to scalding. I often wondered whether it was her way of making sure that my stay was always brief.

'It's just you'd think there must be warning signs,' I said, feeling flushed already. 'Some way of telling that part of Roxi's cycle is about to begin.'

From the opposite end of the bath, Emma raised one eyebrow. 'Like what?'

'Well, she was a bit grumpy with me this morning.'

'Don't go there.' Emma lifted a cautioning finger from the bubbles. 'Just don't.'

'I'm serious,' I said. 'She started headbutting my leg when I was scooping out her breakfast. It was as if I couldn't feed her fast enough.'

'PMT does not make me greedy!' she snapped back. 'Nor does it make me irritable. That just comes from being your wife.'

'We're talking about Roxi,' I reminded her. 'Obviously it's different for minipigs, but you'd think they'd still show signs. I just can't believe one minute she's not brimming and the next she's ready to be served.'

Emma pulled a face. 'Who comes up with these dreadful terms? Those poor lady pigs.'

I could feel myself beginning to cook. Anxious to climb out, and aware that this conversation had veered into a different kind of hot water, I simply remarked that our window of opportunity with Roxi had to occur within the next week.

With Butch and Roxi firmly established in our lives, my body clock had adjusted to minipig time. Frankly, there was no worse way to start the day than to have your dreams invaded by the sound of frenzied snorts and squealing. I also had Roddie to consider. In a bid to be a good neighbour, as well as manage my stress levels, I started to set my alarm to rise shortly before dawn. With the minipigs still sleeping, it meant I didn't have to rush out in a bid to buy their silence with food. Such was my anxiety that eventually I found myself waking minutes before the beeper went off.

Over time, Butch and Roxi really had come to own me. In a way, I felt less like their keeper and more like their bitch.

I didn't mind, so long as I reached the pigsty before the

first bars of birdsong reached their vaulted ears. I could always slip back to bed, after all, and that's exactly what I intended to do the morning after my bathtub conversation with Emma. Before leaving the pigsty, I had pressed down on Roxi's flanks. Once again, she moved away. I left the pigsty and trudged back to the house, the moon still visible in the brightening sky, and then stopped in my tracks. For instead of the steady crunching sound that marked the minipigs' breakfast time, I heard a long, high-pitched and plaintiff whine.

I turned around to see Butch at the feeding bowl, guzzling happily, while Roxi stood behind the gate. Once she had my attention she butted it with her forehead.

'What's up?' I asked her quietly, and returned to the pigsty.

Roxi whined, and hit the gate again. It was clear she wanted to come out. I just couldn't figure out why. Normally, she would be fighting with Butch to eat the most food. This time she looked totally uninterested, and that's when I realised she might be trying to tell me something.

'Emma,' I whispered, back in the bedroom. 'Wake up! We have a live one!'

In the half-light coming through the curtains, my wife looked at me groggily. 'What are you talking about?'

'Roxi!' I smiled. 'She's about to start brimming. I'm sure of it.'

Emma glanced at the clock and sighed to herself. 'Did she respond when you pushed down on her?'

'No,' I said, 'but she wasn't happy about me leaving the pigsty without her.'

'So, what does that mean?'

'Think about it,' I urged her.

'It's not even six in the morning!' hissed Emma. 'The only thing I wish to think about is sleep!'

I realised I was behaving a little excitably here. Then again, I was genuinely elated. This moment had taken a long time to arrive. It had also demanded so much preparation. Consequently, the chance to artificially impregnate a minipig suddenly seemed like an experience akin to sky diving or driving a sports car round a racetrack. I drew breath to calm myself down, and told Emma how Roxi had behaved as I left the pigsty.

'She's looking for a mate,' I revealed. 'Butch is no good to her, after all. She needs a boar that can meet her desires. The hormones must be surging. Roxi is a minipig on the cusp of needing a good service.'

Emma closed her eyes for a moment. 'And that's where you come in, right? The service engineer.'

Her comment wasn't meant as a compliment. Just then, however, nothing was going to put me off.

'I'm still hoping this can be a joint venture,' I said in reply, before heading around to my side of the bed.

I couldn't get back to sleep. In my mind I was ticking off all the things I had to make ready. What I really wanted to do was assess Roxi once again. I only got a chance to do so once Emma had left for work that morning and the kids were safely at school. Having walked the dog and fed those remaining pets that presented no threat when hungry or unexercised, I took myself down to the pigsty.

Once again, Roxi was waiting at the gate. Butch was working on the ever-growing root crater. I felt pity for the poor thing. Not only was he too small to make much of an impression on the ground, his female companion had basically written him off. I figured being infertile meant he didn't give out the necessary hormones to present himself as the boar who could deliver the goods. For personal reasons, it wasn't something I wanted to dwell upon. Besides, I had a simple test to carry out. Manoeuvring behind the minipig, I placed my palms on her rear flanks and applied a little pressure.

For the first time, instead of stepping away from me, Roxi took the weight of my hands. And then, to my delight, she spread her rear trotters apart. Unwilling to look like some kind of minipig tease, I stood back and told her to be ready for a treat at the end of the working day.

'As soon as Emma gets home,' I said, 'it's party time.'

In retrospect, I made a mistake in calling my wife at work to confirm the news. I suppose being at home every day makes you forget that people in offices aren't always in a convenient place to talk. I couldn't possibly have known that Emma was in a meeting when I phoned. I suppose she did sound a little offhand when I asked her if she could pick up some pressed her on on the way home. Especially when I pressed her on what brand would be most suitable.

'Why don't you just text me next time,' she suggested, on her return that evening with a carrier bag from the chemist in hand. 'I was negotiating with a client when you called. Instead of sealing a deal, I spent the rest of that meeting fielding questions about minipigs.' She handed me the bag, hiding a smile. 'I might have found a buyer for one of Roxi's little offspring, though.'

'Really?'

'Apparently his wife and girls have been talking about nothing but minipigs since they saw some on the television.'

'You should ask for a deposit,' I said after a moment, having paused to think this through.

'I can't do that!' Emma said as she kicked off her career heels. 'What kind of message does that give out to a client?'

I shrugged, unable to grasp her point. 'It says you need some commitment surely? What's wrong with that? Our minipiglets will be in hot demand. I guarantee it. If we start doing favours for friends and colleagues, next thing they'll be asking for discounts and that could all eat into the profits.'

Emma gazed at me in a way that made me think my soul was under scrutiny here.

'It isn't just about the money, though,' she said quietly, 'is it?'

Outside, a westward sun was beginning to lengthen shadows. The kids were in the front room, watching something peppered by an awful lot of gunshots and squealing tyres. It was unlikely to be suitable for the little ones, but it did buy us an opportunity.

I had already collected one of the straws from Tom's fridge. It had been a relief when I found he wasn't around. We hadn't fallen out as such, but it did feel like he had withdrawn his support for my bid to impregnate Roxi. All I wanted to do was prove him wrong. He had taught me a great deal about pig keeping. The way I saw things, this was my chance to prove I had the skills to produce a litter of my own. I had even popped to the smallholder's supermarket and purchased a bottle that was suitable for me to fill with defrosted minipig semen. As for Tom's concern about how my family would respond to failure, it just wasn't an issue. Why not? Because I couldn't afford to let them down. With this in mind, I answered Emma in a bid to sound as positive as possible.

'This is about *us*,' I said, and took the lubricant out of the bag, 'and we really should seize the moment before it gets dark.'

'I'm in my work clothes,' replied Emma. 'Is this going to be messy?'

'I'll do the dirty work,' I told her, 'but I really don't think it's going to be an issue. In a way, we're just doing what comes naturally. It's never caused any problems for us, after all.'

Emma drew breath to reply. Then she appeared to abandon whatever she was going to tell me. 'I think I'll wear my wellies,' she said instead. 'It's always wise to take precautions.'

26

Nature and Gravity

I was well prepared. Not only was I in possession of all the equipment, which I kept in a sweet tin left over from Christmas, I had been online and researched how to manage every step of the way. My first task, if I was to maximise our chances of success, was to present myself to Roxi in a way that she would find attractive.

'What *are* you doing?' asked Emma, having just climbed into her boots.

I was standing outside the back door, the sweet tin at my feet, dousing myself liberally in the spray that Tom had recommended.

'It's supposed to make me smell like a boar,' I said, applying several squirts to my neck. 'A hot one, hopefully.'

Emma joined me outside. She walked into the mist still hanging in the air, turned and stepped right out again.

'How can that be attractive?' she asked. 'You smell like a corpse!'

'You're not a minipig,' I reminded her, and collected the tin from the ground. 'Let's see how Roxi reacts when she gets a whiff of me.'

Still keeping her distance, Emma invited me to lead the way.

As well as the bottle containing the contents of the defrosted straw, the lubricant and the long catheter with the spiral tip, my tin also contained a pot of apple segments. I wanted to be sure that Butch and the hens were occupied before we set to work. I also thought Roxi would be more compliant if she was feeding her face. We found her at the gate once more. She looked restless, which I took to be a good sign. From experience, I knew both minipigs would pick up on our presence long before they sighted us. On this occasion, my approach caused Roxi's ears to prick before I'd even let myself through the garden gate.

'Someone's perky,' observed Emma, who had dared to walk alongside me by now.

'I don't think we'll have any difficulties here,' I replied. 'She's ready for me.'

Emma caught my eye. 'You're uncharacteristically confident,' she said, and opened the gate. 'I'm ready to be impressed.'

I was determined to do a good job. Not just for my wife, but to prove a point to Tom as much as myself. It was then that Butch appeared from the shed. Straw hung from his back and his head. He blinked dozily in the dusky light, having evidently been disturbed from an early night in. Given that he didn't have anything better to do, being a boar with no balls, it seemed like an understandable thing to do. I set the tin on the concrete floor and peeled off the lid. I could tell by the way Butch's snout twitched that he knew he was in the company of a virile potentate.

It was time to stand in for his shortcomings.

'This shouldn't take long,' I told Emma, as she scattered apple segments for both minipigs and hens. 'And we both know that's not the first time you've heard me say that.'

She looked across at me and grinned. I thought she would make some cutting comment as I took the catheter in one hand and began to lubricate one end. Instead, she watched me with interest.

'It's weird in a way,' she said eventually. 'Did you ever think you would be doing something like this?'

'Never.' I bent down to examine Roxi's rear quarters. I was as ready as I'd ever be, and so too was the minipig if my observations were correct. I looked up over the length of her back. From the opposite end, my wife continued to observe me closely. Only now she looked a little amused. 'What?' I asked.

'I was just thinking this isn't quite the three way of your fantasies, is it?'

'Emma! Now is no time to be silly.'

I smiled despite myself, which only caused her to giggle.

'I'm sorry,' she said. 'It's just a ridiculous situation.'

'It isn't for Roxi,' I pointed out. 'And I'm at the business end. It's no laughing matter back here.'

Emma took a moment to compose herself. 'You're right,' she said, and scratched the minipig behind one ear. 'You know it means a lot to me. I'm just nervous. Nervous and excited.'

Roxi flicked her tail just then. I was beginning to feel sick with nerves.

'What *do* you see in minipigs?' I asked, aware that what I was looking at from my position behind Roxi wasn't quite what I had in mind. 'It's been chaos since they arrived, and here you are wanting to create more.'

'I don't see it like that.' Emma shrugged. 'I like being surrounded by chaos because there's always calm at the centre of it all, and that's where I like us to be.'

I turned the catheter between my fingers as I thought about this. It would soon contain a solution guaranteed to bring nothing but more upheaval into our lives. Instead of working with two minipigs in my care, I had visions of a small army marauding around me, with several earmarked to stay on a permanent basis. Despite such an unthinkable scenario, I asked Emma if she would be good enough to press down on Roxi's hindquarters. It was too late for second thoughts now. All I could do was get this thing in place, attach the bottle and then think of the money.

'Here goes,' said Emma, and did as I had asked. At once, Roxi shuffled her back legs apart. 'Wow! She doesn't play hard to get.'

'Let's hope it stays that way.' I steeled myself for the next step. 'Nice and steady now . . .'

With a little pressure, the tip of the catheter sunk inside our minipig. I felt like I was doing something awful here, but she didn't seem to mind. By now, all the apple segments had been eaten. Butch and the hens had ventured off in different directions. Roxi, meanwhile, remained quite still. Taking a breath, I prepared to start turning the catheter so the spiral section screwed into place.

'Can I ask something?' said Emma at the same time, which rather ruined the momentum. 'Why is it shaped like that?'

'Ah, now I can answer that,' I said with some pride, for I had read about it on the internet. 'It mimics the shape of a boar's penis, which is designed to lock inside the girl pig.'

Emma crinkled her nose. Despite my less than technical explanation, it felt good to share my knowledge. What's more, I was the one putting all the work in here. It was a hands-on venture that beat slipping off to make the tea any day.

'So,' said Emma, who had thought about this for a moment, 'which way are you supposed to screw it in? Clockwise or anticlockwise?'

As carefully as I could, so Emma didn't notice, I tried it one way, which met with resistance, and then the other, which felt exactly the same. Quietly, I removed the tip and looked at the direction in which the spiral turned.

'Anticlockwise,' I said. 'Trust me,' I know what I'm doing here.'

Emma muttered something under her breath about this being a first. I didn't ask her to explain herself. While attempting to artificially inseminate a minipig, I was well aware of the position I found myself in. Instead, I inserted the catheter once more, this time turning it in the direction it was intended without worrying that I had got it wrong. After a couple of rotations, I noted Roxi's flanks tense and then relax. I smiled to myself. Now came the science bit.

'We're locked in,' I told Emma. 'From here on out, all I need to do is get the goods flowing into the catheter and she will do the rest.'

'How so?' she asked.

'Internal contractions apparently. She'll create a suction and draw it all in.'

'Matt . . .'

I looked up, having retrieved the bottle from the sweet tin. Emma still had her palms placed on Roxi's flanks.

'What is it?'

'That's oversharing. Let's just get on with it, before I change my mind about all this.'

Without further comment, I attached the free end of the catheter to the bottle. I had chosen one with a special connecting cap. As soon as I fixed the tube in place, all I had to do was stand up and hold it high. Then both gravity and nature would run its course.

'There she goes,' I said, as the milky fluid began to snake into the tube. 'Bon voyage!'

Emma turned her attention to Roxi. She remained quite still with her head bowed and her eyes half shut.

'I think she's getting bored,' said Emma.

I didn't like to ask how she had drawn this conclusion. I just focused on the fact that the boar semen was on its way.

'Even if she isn't into it,' I said, 'from where I'm standing it's fascinating. I could do this all day!'

I glanced at Emma when she didn't respond. At first I thought she was staring at me. Then I realised her focus was fixed on a point a little way behind me. I turned around, and found our four children lined up behind the fence. To my eye, they looked like witnesses to some terrible atrocity.

'Eeew.' This was Lou, who looked very pale indeed. 'What are you doing, Dad?'

'It looks like it hurts,' chimed May. 'You're hurting our minipig!'

'No,' I said, matter-of-factly. 'This is how we make minipiglets.'

Lou tugged her phone from her pocket. 'This I have to film,' she said, flipping it open. 'It's guaranteed to go viral.'

'No, you don't!' I outstretched my palm. 'This is not for the internet, Lou!'

Honey and Frank looked baffled at all this, but watched intently. As the boar semen neared its target, I realised I had possibly made life difficult with regard to our children's sex education. It was one thing to catch your parents at it. Finding them assisting a family pet wasn't something they were likely to forget for years to come.

'Go back inside,' said Emma, but her words fell on deaf ears.

Glancing at the catheter once again, I realised the boar semen was close to reaching its target. With no time now to manage the children, I figured all I could do was involve them in this special moment.

'Just a couple of seconds to go,' I said, aware that Roxi's contractions would take over just as soon as the fluid in the catheter passed inside her. 'Five . . . four . . . three . . . two . . .'

I faced my offspring to mark the moment of contact, recoiling just a little when Frank pressed himself against the mesh for a closer look. Maybe it startled Roxi, too, or perhaps she interpreted my countdown as some sort of starting signal. Either way, she moved off so abruptly that the catheter popped from the bottle and trailed on the ground behind her.

'Now you've broken her!' declared May, as I raced in vain to grab the tail end of the tube. 'Dad has *broken* our minipig!'

Twice I swooped for the leaking catheter. By then, however, Roxi had decided that her interests now lay in hunting down any stray bits of apple segment that had dropped from Butch's jaws. She found some on the slope of the crater. I watched her fall upon it, and stand there chomping while every last drop of precious boar semen pooled from the tube into the soil behind her. I looked to Emma. From her expression I knew she felt this was my failing.

'We have learned something here,' I said, determined not to consider this a defeat. 'It's too dark to try again now, but tomorrow we shall do it without an audience. Evidently minipigs aren't comfortable being watched.'

'That's fine by me,' replied Emma, brushing specks of dirt from her business suit. 'As soon as the kids are at school, you're free to give it another shot all on your own.'

I had two attempts left. The way I looked at things, having marked down the first go as a practice run, the odds remained in my favour.

That evening, I felt quite optimistic as I washed, sterilised and dried both catheter and bottle. Roxi hadn't misbehaved that badly. She was content so long as she had something to eat in front of her. It really was all in the preparation, and that included defrosting another batch of boar semen. So, following the school run the next morning, I pulled in at the entrance to Tom's smallholding. My heart sank a little on seeing his Land Rover down by the stable. Barry had advised me to use all three straws, to maximise the chances of conception. I just couldn't do much about the look on my face. It was bound to tell Tom things weren't going as smoothly as I hoped. Still, I couldn't let that stop me. Creating minipiglets had become my mission in life.

'How is it going up there?' he asked, when I found him mucking out a stable.

'Good,' I replied, facing him from the door with my hands in my pockets. Sunshine streamed in from behind me. 'Couldn't be better.'

Tom continued to rake at the floor. 'Help yourself to another straw,' he said. 'Assuming that's why you're here.'

I tried to appear surprised that Tom should even suggest such a thing. 'I guess I could do that,' I said, 'to be on the safe side.'

'Go for it. And good luck.'

'I won't be needing luck,' I told him, and turned to head for the workshop.

'Wait a moment.'

I turned around. Tom was facing me now, leaning on the rake handle.

'What is it?'

'If you need help, you know, despite everything I'm happy to—'

'Thanks,' I cut in, 'but I'm fine just as I am.'

* * *

Back home, I placed not one straw but two into a bowl of warm water. My exchange with Tom had persuaded me to take both. I had allowed my pride to get the better of me, and that left me feeling awkward. I really couldn't face seeing him again until I had good news. All I could think was that success in impregnating Roxi would somehow smooth things between us.

This time around, only Sesi watched me douse myself in boar spray. Lying out in the yard, she showed little enthusiasm for what I was about to do. Roxi, on the other hand, evidently intoxicated by my aroma, couldn't wait to greet me. She nuzzled the back of my legs as I let myself through the gates, and made nothing but happy noises.

'Easy now,' I said, as Butch and the chickens crowded around me. 'This is between Roxi and me.'

Reaching into my pocket, I grabbed at a stash of raisins I had packed for this moment and then flung them into the crater. Such was the diameter and depth it was beginning to look like an open-cast mining operation, but at least it kept them busy. As Butch and the hens gave chase, Roxi looked torn. Her instinct was to track down every last available offering and gobble them up, but this minipig was brimming. For three days, the reason for her existence was entirely rewired. Knowing Roxi, however, and the possibility that the lure of food would prove too powerful for her to resist, I dropped another handful of raisins right under her snout. That way, while Butch and the hens continued to scour the soil for treats, Roxi was free to satisfy her needs without moving from the spot.

'OK, baby,' I said soothingly. 'This time, we're going all the way.'

Once again, I went through the same steps as I had the day

before. I found that Emma had been right. I didn't need help. Instead of relying on a spare pair of hands to press down on Roxi's hindquarters, I summoned the same result by draping our groundsheet over her back. It wasn't dignified, but the weight of it worked. In fact, the steps that followed went without a hitch. I inserted the catheter, locked the bottle onto the free end and lifted it high. Roxi behaved perfectly. She didn't move a muscle. Once again, the fluid began to crawl into the tube.

'Steady now, girl,' I said soothingly. 'It'll be over before you know it.'

'Mr Whyman. There you are!'

I recognised my neighbour's voice, calling me from the garden gate, I just wasn't prepared to find myself in company. I turned, without thinking, which served to tug the catheter from the bottle. Sensing the sudden detachment, Roxi grunted in disappointment.

'Roddie. I'll be with you in just a moment.'

'You look busy,' he said. 'Stay there, I'll come to you.'

This wasn't part of the plan. The minipig might have been waiting for me to reattach the tube, but somehow that didn't seem appropriate. Just then, Roddie let himself into the garden. It was then I realised I was still holding the bottle aloft. I glanced up, saw the last of the liquid dripping out, and promptly hid it behind my back as my neighbour reached the gate.

'Just hanging with the minipigs,' I said, as if somehow I needed to provide him with an excuse. 'It's good to spend some time together.'

'You certainly have a bond,' he said, sounding warmer than usual. 'I do not appreciate being awoken by her screams first thing in the morning, but just now I heard you talking sweetly to her. Clearly she craves your company.'

'Oh, we've had some memorable times together,' I replied.

With my free hand, I reached back and tugged the groundsheet from Roxi's rear haunches. With a grumble, sensing I had just let her down very badly indeed, she sauntered off to join Butch and the hens. It was then I realised the catheter was trailing behind her once more. In a bid to block Roddie's view, I spun around to face him. I couldn't bring myself to explain why I felt the need to hide what I'd been doing. It wasn't illegal, after all, and yet I felt a strong sense of embarrassment. It was as if I'd been caught undertaking something foolish and reckless; another pet expansion programme within a household that already boasted enough animals to open a petting zoo.

Fortunately, Roddie didn't seem to have noticed the tube. My relief was short-lived, however. Having filled his field of vision, my neighbour could only look me up and down. Weirdly, he didn't then return his gaze to mine. Instead, he just stared at my boots. I followed his line of sight. What I saw momentarily shaped my expression into shock and horror. I had barely recovered when he dragged his attention back to my face. Roddie looked pale. I wanted to explain myself. I just worried it would make things worse.

'Mr Whyman,' he said, struggling to find his voice. For a moment he looked like he had forgotten the purpose of his visit. When it came to him, he addressed me while evidently battling with himself not to take another look at my boots. 'I'm here about the village fete,' he pressed on. 'The committee were hoping you could help out on the face-painting stall once more.'

For the last two years, I had been roped into this role. Unlike the other volunteers, I possessed no discernable talent at transforming children into lions, fairies and butterflies. Instead, by way of a cover, I specialised in doing 'army camouflage'. It was boring to paint, but impossible to screw up. It meant happy little soldiers and parents who didn't look at me like I might go on to chance my hand at dentistry or piloting a passenger plane.

'I'd be delighted,' I told Roddie, while furtively attempting to wipe my boot on the back of my jeans.

'The committee will be pleased,' he said somewhat absently.

This time, my neighbour couldn't resist glancing down again. For a village elder, one so buttoned up in every way, he looked fascinated and appalled in equal measure.

'It's not what you think,' I said, when Roddie finally realised he was staring once more.

'I assumed it was minipig semen.' He took a step back from me. 'Am I mistaken?'

27

Third Time Lucky

This was the last straw. Not just metaphorically, either. I really did have one more shot at artificially impregnating Roxi, and only one more day in which to get it right.

'How did you get on?' asked Emma after work. 'Are we *in pig*?'

My wife had done some research, which was a bad sign. I knew what she meant. In my preparations, I had come across the term myself. The fact that she knew the correct expression for a pig expecting piglets didn't trouble me. I just knew from experience that it meant she was about to pay much closer attention to the venture.

'It went as well as can be expected,' I assured her. 'Fingers crossed, eh?'

As ever, the little ones were crowding around her to reel off the high points of their day. Emma crouched and listened to them both. I thought about how she would react if I took my turn and shared the most notable thing that had happened to me. It was only a passing thought. One that left me determined to make the third straw count. I just couldn't bring myself to prepare her for the possibility that we might not be having minipiglets after all. I wasn't even able to turn to Tom for advice. He'd already offered me a piece of his mind, and I had chosen to ignore it. I was trying to do a good thing here for the family. So far, all it had brought me was grief and self-doubt. Just then, Emma looked up at me.

'It's a beautiful evening,' she said. 'Why don't we let Butch and Roxi into the garden for a play?'

'Because they'll turn it upside down,' I said, as if I needed to spell that out. 'It's what they're programmed to do.'

Smiling, with the little ones each holding a hand, Emma rose to her feet.

'Just once,' she said. 'We'll make sure they don't get up to too much mischief. It would be the perfect end to a long day.'

Outside, at the pigsty gate, Butch and the chickens greeted us eagerly. Roxi was stationed at the rim of the crater. She was crunching on something. It was a troubling sound, like bones fragmenting, but I knew for a fact that Misty the cat was safely entombed.

'What is that noise?' asked Lou, who had elected to join us, as had our second born.

'If it's a bird, I'll cry,' said May, clutching her elbows protectively.

Just then, Roxi bit down successfully on whatever was in her jaws. Little clouds of red dust puffed from the sides of her mouth, which told me what it was.

'It's OK,' I said, to reassure May. 'It's just a brick.'

'A what?' Emma faced me side on.

'There are a few buried around here. She's just unearthed one.'

'To *eat*?' Emma looked flabbergasted. 'I think it's time Roxi got out more.'

Still munching on her prize, dribbling what was basically rubble, our female minipig heard the gate squeak as I opened it. Immediately, her ears pricked and she peered around. I stepped aside, as did Emma and the kids.

'Here's your chance,' I said, gesturing at the garden. '*Fly*, my pretties!'

It was the chickens who ventured out first. They did so tentatively, their heads twitching one way and then the other, as if to stake out the area in preparation for Butch and Roxi. For a short while both minipigs remained at the threshold, much to May's disapproval.

'Look what you've done!' she said accusingly. 'It's like they've got Stockholm Syndrome.'

'Oh, hardly!' I glanced at Emma, still wondering if this was a sensible thing to be doing. 'They just can't believe their luck.'

'Give them a chance,' she said, crouching to coax them out. 'Maybe Roxi wants to rest because she's only just conceived.'

I closed my eyes for a moment. Silently, I reminded myself that I still had one last opportunity to make this dream come true. When I opened my eyes, I did so with a gasp of surprise.

'That was my foot!' I cried out, hopping on the spot.

Having just trampled on it on her way out, Roxi was now midway across the garden, sweeping the grass with her snout. Butch had fanned away towards the rabbit runs. Inside, several bunnies sat up on their hind legs and looked very startled indeed. I missed the moment that Roxi rooted her first strip of turf. I was too busy watching the chickens scratch up the flower bed. On hearing the kids squealing as much as the minipigs, I turned and saw her hurl the extracted sod over her back. I moved to intervene, only for Emma to grasp my wrist.

'Let's leave them,' she said, and laughed as the little ones chased after Butch. 'Everyone is having fun.'

Even the dog was drawn by the commotion. Barking excitedly, she raced into the garden and set about circling anything that moved. As she swept past the chickens they scattered in a panic, only to regroup, looking set for retribution. Fortunately, May stepped in to diffuse the situation with a handful of seeds from the bird feeder. Meanwhile Lou had flipped open her phone. She was busy filming proceedings, but couldn't seem to decide who was causing the most chaos. She kept switching from dog to minipig to poultry. I imagine the result would look as giddy and unsettled as the kind of footage you see when gunfire breaks out over a crowd of protestors. Only Miso was missing from this mix. I spotted him on the window ledge in the front room. He was looking out with the same thousand-yard stare that had haunted him since the minipigs arrived.

'Isn't it about time the cat came off the tranquillisers?' I asked, having pointed him out to Emma. 'Butch and Roxi have been here for some time. They're part of the family now.'

In response, Emma linked her arm with mine. She found my hand and squeezed it. 'I'm glad you said that.'

'What? About the cat?'

Emma glanced at me. 'Well, you're right about Miso, but I mean about the minipigs. I'm pleased you consider them to be a part of the family.'

'Even though one has developed a taste for house bricks?'

Emma smiled to herself, all the time watching children and animals scamper this way and that. The sun was beginning to set. It was that brief time of day when the light turned golden, shadows stretched to their full extent and tiny flying bugs whisked overhead.

'You know what?' she continued after a moment. 'Whenever work gets on top of me, I *always* look forward to coming home. You know I'd rather be here all the time, but it's little things like this that get me through the day.'

As we spoke, Roxi continued on her spree of garden anarchy. Intoxicated by the smell of fresh grass, she was rushing around trying to tear off as much as she could while ducking from the dog, Frank and Honey. It was as if she couldn't quite believe her luck, and was just trying to cram as much as she could inside her mouth before I sent her back. Even Butch had come alive. I watched him pull a plant out of the border and shake the roots free from soil. I should've been appalled. Instead, surrounded by noise and chaos, I felt quite at peace.

I elected to cook supper that evening. It minimised the risk of Emma opening up the freezer compartment and me subsequently sleeping on the sofa. When I came to collect the final straw the next day, I did so with a sense of relief. I hadn't enjoyed this process as much as I'd thought I might. I had set out on the road to artificial insemination feeling fired up and optimistic. It was something I had never anticipated I would do, but as a challenge it seemed irresistible. In practice, it had proven to be a frustrating business. Messy, too.

As I prepared the equipment for one final try, I told myself this had to be the hardest part of the process. Admittedly, giving birth probably wasn't going to be much fun for Roxi, which was when I realised I might be required as a minipig midwife. The very thought did little for my spirits. Looking at the dog, watching me from her bed, her expression summed up how I felt.

'I take it you won't be joining me?' I asked her, on making my way to the back door. 'Oh, come on, Sesi! Show some support, eh?'

The dog didn't move. As I reached for the boar spray, she just gazed up at me with what seemed like pity in her eyes.

Maybe the novelty had begun to wear off for Roxi as well.

As I approached, reeking of all the right pheromones, she didn't seem so wildly excited to see me. She met me at the gate once more, but appeared more interested in the apple segments in my pocket than getting down to business.

'One more try,' I said, and set down my box of bits. 'Third time lucky, right?'

With the segments distributed in order to keep everyone happy, I went through the first stages of the process once more. Roxi knew what to expect. She didn't try to resist, but then nor did she seem particularly willing. Eventually, with the catheter in place, I stood tall for a moment just to rest my back. As I did so, Roxi's huge ears caught my attention. They really were remarkably long, out of whack with the rest of her body, and bounced as she gorged on her treats. There was no escaping the truth, I thought to myself, this really was one ugly minipig. It made me wonder what her offspring might look like.

Carefully, as if to answer my question, I picked out the bottle from the box. This was premium minipig semen, so Barry had said, and I had no reason to disbelieve him. Nevertheless, with so many generations of cross breeding going into one minipig, the shape and size of the litter I could create here might come as a complete surprise. Not just to me but those people who would pay out their hard-earned money to take a minipiglet home with them in a cat basket. At knee-height, Roxi was certainly on the small side for a pig. It's just she stretched the definition of mini, weighed more than I did, and was literally capable of eating the house. As she'd just bolted down the last apple segment, time was running out for me.

So too, I realised, was any enthusiasm I once had to see the job through.

Roxi turned to me expectantly, but I was all out of treats. The way I saw things all of a sudden, I had nothing more to give. We had two fine, patience-testing, neighbour-baiting minipigs with an appetite for destruction, both of whom meant a great deal to us. Why would we want to distract from that by creating any more?

Twisting off the lid from the bottle, and without further thought, I tipped the contents away.

'It's not you,' I told her. 'It's me.'

Roxi turned her head. She batted her lashes, looking a little confused. With a practiced hand, I removed the catheter for the final time. My heart just wasn't in this venture any longer. It was one of those decisions that just popped into my head, but made complete sense to me. I had no doubt that I was doing the right thing. After cranking hopes as high as they could go, I just wasn't sure how I would break it to the family.

The straws cost a lot of money. I had no intention of admitting I poured one into the grass. Instead, when Emma asked me how I had got on, I made positive noises.

'That's great!' she beamed, and even clapped her hands in delight. 'I really admire your dedication, Matt. Giving it three goes *must* mean Roxi is in pig. Just imagine, three months, three weeks and three days from now we'll have our very own litter!'

I forced a smile. It was as fake as I felt. I also realised that Emma wouldn't have to wait that long before learning the truth. In exactly three weeks, Roxi would be brimming once again. We both knew that was effectively the pregnancy test for pigs. If she was expecting, she wouldn't take kindly to anyone pushing down on her rear haunches. As soon as Emma laid her hands on Roxi, her dream would come to a sudden end. Of course, I could suggest that perhaps she had a problem when it came to conceiving. I just knew that blaming the minipig for Emma's disappointment would leave me feeling even worse about things.

And so, faced with a dilemma that could have no happy outcome, I chose to pretend it wasn't happening.

With every day that passed, the opportunity to tell the truth became more distant. Looking back, I think I must have become somewhat detached. It didn't feel right to encourage any kind of conversation about rearing minipiglets. Whenever the subject arose, between Emma and the children, I would make myself busy elsewhere. Nobody seemed to pick up on my reluctance to get involved. With so many animals to care for, it was easy to slip away when things became uncomfortable.

I know that from an outside point of view this wasn't an issue that by rights should provoke a personal crisis. Making minipigs isn't high up on the list of things to keep a grown man awake at night. Nevertheless, I knew that for Emma it represented so much more. This was her dream family in the making. Something she had longed for as a little girl left to her own devices.

With a week to go before Roxi revealed all, the strain began to show. This was made apparent to me at the village fete. I was an hour into my afternoon shift on the face-painting stall. A lot of kids had shown up, but my services had yet to be called upon. My two fellow face-painters were mothers I knew from the school playground. Judging by their way with the brushes, it looked like they spent their entire time between drop off and pick up in front of a canvas and easel. In a bid to make myself look busy, and also to up my game from *Basic Camouflage for Boys*, I had painstakingly decorated my own face. I was just putting the finishing touches to the design when Tom sauntered into my field of vision. I nodded at him before finishing the job. He smiled, nodded back, and then stood there with his hands in his pockets, appraising me it seemed.

'You look troubled,' he said eventually. 'A troubled Smurf.'

'I'm not a Smurf,' I said pointedly, and completed the last gold freckle. 'I'm a warrior from the Na'avi tribe. From *Avatar*. The film?'

Furrowing his brow, Tom appeared to consult his memory. 'Smurfs have blue faces,' he said. 'You can't do blue and be anything but a Smurf.'

Despite our recent differences, I failed to hide my smile.

'I can paint your face, if you like? One pound fifty a go. It's for the village hall fund.'

Tom delved into his pocket. 'Let me just pay you the money,' he said, and flipped several coins into the pot on the table. 'An oversized Smurf might scare the kiddiewinks.'

Tom wasn't really doing justice to my work here. Yes, the primary colour was blue, but I had also detailed the subtle tribal markings across my forehead. Back home, for reference, I had found a picture from the movie online and printed it off. In the absence of yellow contact lenses, to complete the alien look, I had applied gold paint to my eyelids and then dotted my cheeks with glitter. It looked pretty good, I thought. The only thing missing were the minipig-like ears, but I didn't want to think about that. Tom, however, had other ideas.

'So,' he said, evidently keen to strike up a conversation. 'It can't be long before you find out if Roxi's in pig.'

'Seven days from now,' I told him, cleaning my brushes at the same time. 'I'm keeping an open mind.'

When I looked up, I found Tom observing me closely.

'It didn't work out,' he said, 'did it?'

I drew breath to reply that we'd soon know one way or the other, only to sigh to myself. I looked around. Not only was there little sign that my publicity attempt could draw a crowd, Emma was out of earshot. She was over on the bookstall, no doubt preparing to drive a hard bargain on dog-eared Seventies paperbacks. With Lou and May occupying the little ones on the bouncy castle, it meant I was free to make a much-needed confession.

'It was never going to happen,' I admitted. 'I bungled the first two straws. The third one I just chucked away.'

Tom listened to this and then sat down in the plastic garden chair facing me. He leaned forward on his elbows, his great hands clasped together.

'I'm assuming Emma doesn't know this,' he said.

'She's never even come to terms with my decision to have a vasectomy,' I said. 'If I admit to wasting a perfectly good batch of minipig semen, I could be sleeping in the shed on a permanent basis.'

Tom nodded to himself when I said this, as if I had reminded him of something else.

'You might like to check that car battery in yours is full of juice,' he told me. 'The electric fence needs to be crackling if you want to keep the pigsty nice and secure. Your neighbour wasn't wrong. There are definitely thieves about.'

'Have you seen them, too?'

'Oh, I'm sure I'll catch up with them,' Tom assured me, sounding very casual about the matter. 'I left my Land Rover unlocked outside the house one lunchtime recently. Came out to find Rosie had been stolen from the back.'

'Rosie?' I looked at him with concern. 'Who is Rosie?'

'My angle grinder,' he said. 'Rosie and I go way back. She cuts through Indian jade stone like a knife through butter. I can't lay another terrace until she's back where she belongs.'

Only Tom could have fallen in love with an item of heavy machinery. Nevertheless, Rosie's disappearance left me fretting that Roxi and Butch might be next.

'I'll make sure the battery is charged,' I said, and promptly paused for a moment. 'How do I charge it?'

Tom shook his head in apparent disbelief. 'You can leave it to me,' he offered. 'Better safe than sorry.'

By now, several small children had noticed the Na'avi warrior sitting at the tent. I noted one of them tugging at his mother's top and then pointing at me.

'I had better get to work,' I said.

Tom glanced over his shoulder. He acknowledged the mother now opening her purse and climbed from the seat.

'That's what I like about Smurfs,' he said, stretching now. 'A strong work ethic.'

I chose to ignore the comment. Instead, as he turned to leave, I had one last thing to say.

'Tom.' I waited for him to face me once more. 'I just want to let you know that I'm sorry. I should've listened to your reservations. The whole breeding thing has been horrible.'

'Oh, I wouldn't go that far,' he said. 'What you've done took courage. We both know minipig semen isn't cheap, but you know what? I'm sure Emma will come to respect your reasons . . . in time.'

I didn't want to think about how long that would be, or what I would have to suffer until my wife came to terms with my actions.

'Above all,' I told him, 'I did it for the family. More minipigs would break us.'

'I know that.' Tom stepped back out into the sunshine. 'But two could be the making of you all.'

I had no time to reflect on this. As soon as he turned away, so my first customer approached.

'Take a seat!' I said cheerily. 'What can I do for you?'

The boy settled without once taking his eyes from me. In the presence of an alien hunter, he seemed in awe and entranced. As several more children headed to form a queue behind him, I wondered if I had enough blue paint to last the afternoon. Without taking his eyes off my face, the boy handed me his money.

'I'd like a camouflage face, please. The same as last year.'

When we first moved to the village, the annual fete felt like a throwback to my childhood. With faded flags and bunting flapping in the breeze, it reminded me of being a kid again. The bouncy castle was a bit more sophisticated, but everything else, from the gurning competition to the coconut shy and the tug o' war was much the same. As an adult, though, all the magic was missing, and this year was a case in point. While I was painting yet another camouflaged face, Roddie dropped by with a cardboard fruit box in his hands. It contained a jar of mayonnaise, a wilted pot plant and a book of raffle tickets.

'Mr Whyman, may I ask you something?' I took a break from daubing my young customer's face with a sponge dipped in green paint. The contents of the basket were frankly pathetic, not that I would dare to say such a thing to this pillar of our community.

'Is this the winning prize?' I asked instead. 'I can offer a free face-painting masterclass if that helps make it seem a bit more attractive. I specialise in alien faces now.'

Roddie held the box protectively to his chest. 'The prize is a trip in a hot air balloon,' he said. 'These are my donations for the tombola stall.'

'Of course.' Quickly I returned to painting the cheeks of the child in front of me. 'So, how can I help?'

'I think we both know,' he said, in such a way that I felt obliged to return my attention to him. 'Your pigs aren't getting any smaller, Mr Whyman. I have chosen not to speak up about the noise in the morning because we are people of the countryside. Nevertheless, the volume is a nuisance. Lately, it's like being awoken by the sound of a pneumatic drill.'

'I know,' I said with some sympathy, before reminding myself that Roddie hadn't come here to share the pain. 'And I'm sorry,' I added. 'Roxi has been unsettled recently, but she's sure to settle down soon. Despite everything, we've decided not to breed her.' I adopted the same expression the vet had offered from my doorstep when he knocked to break the news about Misty. 'It's for the best,' I added. 'Sometimes, things are meant to be.'

'Two are indeed a nuisance,' he said. 'A whole litter would be unacceptable.'

'Well, I can assure you that isn't going to happen.'

As I said this, Emma and the little ones loomed into view behind him. Frank and Honey were in tears, which was enough to persuade Roddie to move on. Frank was rubbing the side of his head. Honey was nursing her elbow. I didn't need to ask whether they had just come off the bouncy castle. The only thing I couldn't be sure about was whether Emma had just overheard me. Judging by the frown she offered when our eyes met, I worried at first that she had registered every word.

'What have you done?' she asked, which caused me to shuffle in my seat.

It was only when she stepped in under the awning that I realised she was appraising my face.

'Can't you tell?' We had watched *Avatar* together only recently, but Emma just looked blank. 'Don't say the Smurf word,' I added. 'I'm sensitive about that.'

Emma smiled, but only briefly because the little ones had just started blaming one another for the bouncy castle collision.

'They're hungry,' she said. 'I'm about to queue for burgers. Can I get you anything to eat or drink?'

I glanced at my watch, and told her I would pass. Unusually for me, I wasn't face-painting with a warm pint of lager beside me from the beer tent. This was because I knew that I would have to drive up the hill to feed the minipigs their tea, and that time was beckoning. Usually, at home, Butch and Roxi would let me know that mealtime had arrived by screaming. If I was late, I could expect to hear them as I climbed the lane. As Emma and the little ones headed off to find the burger stall, I hurried to finish camouflaging the face of my young customer. It wasn't difficult. I could've done it with my eyes closed, and indeed when he looked in the mirror that's pretty much how it appeared.

'You're good to go, soldier,' I said, before he had a chance to complain. 'And if you'll excuse me, I have a mission of my own to pursue.'

I didn't just give the beer a miss because I had to drive. At the fete the previous year, as the sun beat down, I had arguably enjoyed a pint too many. This became apparent when a pushy lad sat down and demanded that I paint an England flag on his face. One that didn't end up 'wonky', as he so politely put it. In response, I had dipped a thick brush in black and proceeded to draw a pair of spectacles around his eyes, followed by a little moustache. Fortunately, I was on nodding terms with his father, and so no offence was taken. Nevertheless, in the sober light of day, I realised I might've pushed my luck.

I reflected on this as I followed the lane up the hill. It made me smile. With the window down, approaching the house, it was a relief not to hear the minipigs kicking off. Even though I knew Roddie was at the fete, it just wasn't a pleasant noise. Maybe, I thought to myself, on pulling into the drive, Butch and Roxi were learning not to be so antisocial.

As well as feeding the minipigs, I also needed to let the dog out to relieve herself. I would take her for a walk later in the day, when the heat had left the air. As I approached the back door I could hear Sesi whining. I didn't think I had left her for that long, but evidently she was busting. On opening the door, she practically knocked me to the floor in her bid to get out.

'Calm down!' I called after her as she raced down the length of the yard. 'Sesi! Relax!'

I found her with her front paws resting on the top of the gate to the garden. She was bouncing on her hind legs, scrambling in vain to get over the top. Tutting to myself, I ordered her off and opened up. The dog raced out, but didn't head for her usual spot in front of the shed. Instead, she sprinted for the pigsty.

I wanted to do likewise when I saw the reason why. Instead, shocked by what was in front of me, all I could think was that something very bad had happened in my absence.

28

One of Our Minipigs is Missing

Inside the pigsty, behind the crater, the fence panel had toppled away. More immediately, Butch was dozing on top of it. For one horrible moment, I thought he was dead. Then I realised he was breathing, and just sprawled out in the sun.

Roxi, however, was nowhere to be seen.

With my heart hammering, my first thought was that she had been stolen.

'Where are you?' I asked, mostly to myself. 'What am I going to do?'

Beyond the fallen fence, I could at least see the chickens. All four were poking about in the grass under Roddie's apple trees. Sesi was at the gate to the pigsty, whining to be let in. My dog had many downsides, but one advantage was the fact that she offered reassurance. At night, both Emma and I felt better knowing that she was at the back of the house, and just then it was good to have her at my side. Sesi also had a nose like no other dog. If there was a trail to follow, she would find it. I moved to let her into the pigsty, only to freeze on hearing something familiar, if a little distant for my liking.

'*Roxi*!' My relief didn't last long. For there was something a bit wrong about the noise she was making. It was definitely coming from my neighbour's garden. What troubled me was the angry if slightly erratic nature of her grunts and squeals.

Grasping Sesi by the collar, I crouched before Butch. He didn't stir. I placed my hand upon his side. Just to be sure he was still with us. Without doubt, this minipig was simply sound asleep. It wasn't just his breathing that confirmed this, but the fact that he was snoring. I glanced at the dog. 'What is going on here?'

Sesi seemed most concerned. She sniffed Butch eagerly, and then licked him under the snout. It was then I noticed that his whiskers were coated in a sticky-looking, fibrous substance. I touched some with my fingertip and held it under my nose. The rich, fermented, cider-like smell was unmistakeable. Duly I returned my attention to the orchard. Littered under the trees where the chickens were scratching, I noted the gorged remains of many apples. Most looked soft and brown in places. If the smell of the pigsty had masked the fruity aroma from me over recent weeks, to a pig it would've been irresistible. It told me everything I needed to know.

Butch was out for the count because he was drunk.

I could only think that Roxi was also under the influence. Where she differed was in her response to the alcohol content of the apples she had eaten. In a way, I was relieved. Nobody had stolen the minipigs, it seemed. In their relentless excavation of the crater in the enclosure, they had simply undermined the posts in order to get into the orchard.

Butch had behaved like a typical male. He'd basically got himself utterly hammered and then passed out. Stepping over the fence, still clutching Sesi by the collar, I only hoped Roxi wasn't the kind of inebriate who would spoil for a fight. We all behave differently after a few drinks. Personally, I am moved to paint comedy glasses on small children. I certainly wasn't the type to go looking for a conflict. I couldn't ignore the note of fury in her grunting, however, which is why I brought the dog along. Ducking under the branches of the apple trees, I also had to brace myself for the distinct possibility that Roddie's immaculate lawn might not be in its finest state. The first random furrows in the grass confirmed my fears, but that wasn't what seized my attention. It was the van, parked in the entrance to his drive. The same one I had spied prowling the lane at night.

The vehicle had been reversed in from the road. The back doors were wide open. I could see bits of machinery inside; stacked up at the back and half covered by a tarpaulin. I crouched beside Sesi, aware that she was straining under my grip. I didn't want to release her. Not until I knew what was going on. Roxi could still be heard grunting. She was just nowhere to be seen. Even so, she had left a trail to follow. Not one that relied upon scent, I should say. Looking towards Roddie's house, the minipig had basically torn up a path through the lawn. It scissored one way and the other, before disappearing around a bank of rhododendrons that screened the yard behind Roddie's house. I realised I had my mobile phone with me. My first thought was to call the police. If I'm honest, the reason I opted for another number was because I worried that I might also have crossed the law myself. Roxi was on Roddie's premises without a movement license, after all.

'Tom,' I whispered, when he picked up. 'I need your help like never before.'

I was familiar with things being out of control but this went beyond what I could manage. Nevertheless, my thoughts remained with Roxi. I couldn't just stay out of sight and wait for Tom. I had to act, and fast.

'Sesi,' I whispered. 'Don't kill anyone, OK? Just look fierce. Nothing more.'

Unfortunately, my dog was no Lassie. She didn't understand what I was talking about. Nor could I trust her to do the right thing. With my adrenalin pumping, and a White Canadian Shepherd in my grasp, I left the cover of the orchard and hurried towards the rhododendrons. Closing in, I realised it wasn't just Roxi who was making a noise. I could hear other activity as well. On reaching the bushes, I crouched once more and peered through the foliage.

Sure enough, there was Roxi. She was rooting frenziedly at the lawn, just in front of the patio. The damage to my neighbour's garden might've been extensive, but it was the two figures in the yard that caused me to freeze. They were hunched over a lawnmower, which Roddie stored in the shed back there. The mower was his pride and joy. A top-of-the-range model, one he cleaned and polished after every use, I very much doubted he would let anyone borrow it. As one of the men checked it over, the other turned back for the shed. When he reappeared, he did so with an armful of gardening tools. Putting aside the fact that there probably wouldn't be much call for any of it thanks to Roxi's path of destruction, I was certain these men had no business being here. Roxi, meanwhile, had picked up on the presence of Sesi and me. I knew this because she stopped digging all of a sudden, and bellowed at us. It looked as if she expected me to march her back to the pigsty, and was voicing her displeasure at the prospect. Having excavated what could've served as a shallow grave, this was one minipig preparing to stand her ground. She was evidently intoxicated. It wasn't just the way she swayed in a bid to stay upright. Glancing at what was in the hole, I realised that the undiluted rage she now directed at us had previously been focused on a partially exposed soakaway pipe. One that had possessed the audacity not to get out of her way.

As I feared, Roxi was shaping up to be a violent drunk.

She dropped her head just then, glaring at me still, only to lose her footing on the slope of her dig as she attempted to steady herself. The thud that accompanied her collapse into the hole was enough to draw the attention of the two men by the shed. They must've seen Roxi at work. I could only think they had no interest in her. Only now, as the minipig struggled to her feet once more, I realised they were looking straight in my direction. It was all too much for Sesi. She barked, just once, in what sounded like excitement.

'Shhh!' I hissed, but it was too late. The pair abandoned their haul and stepped out onto the patio. They looked much younger than I had assumed, late teens perhaps. One was overweight, but with a broad chest, pumped arms and a lizard tattooed on his neck that I doubted anyone criticised to his face. As for his smaller accomplice, he was wearing jeans that badly needed hitching around his waist. He seemed very self-assured as he looked around, with a provocative glint in his eye. The dog barked once again. All I could think to do was let her go in the hope that she would intimidate them. Taking my hand from her collar, Sesi leaped out across the grass. She ignored Roxi completely. As the dog barrelled around her, the young men looked shocked and startled. One cursed. The other turned as if preparing to run from this white wolf, as I had hoped. Only Sesi didn't pounce upon them. She didn't even growl. She just skidded on the paving and circled them with her great tail wagging. The pair looked confused, but only for a moment. When one of them attempted to shoo her away, my dog just treated it like a game. Then the other one tried to kick her, and that's when I knew that I had to act.

'That's *enough*!' I yelled, and stepped out from the foliage.

'Don't you dare harm my dog . . . or my minipig,' I added, as Roxi joined the two men in just staring at me.

I surprised myself by how aggressive I had just sounded. I certainly couldn't back it up physically. The two men glanced at one another, and then returned their attention to me. It was the larger one who was first to chuckle to himself.

'What are you going to do?' he asked. 'Shoot us with a poison-tipped arrow?'

His comment confused me, until I remembered my painted face.

'Great film,' his accomplice said, seemingly unconcerned by my presence. 'A bit long in places, but the effects were amazing.'

'This is private property,' I warned them, feeling weak at the knees and slightly stupid with my blue skin and yellow eye shadow. I had completely forgotten about it, in fact, until now. Still, this was no time to be pleased that someone had actually identified who I was supposed to be. Instead, I jabbed a finger at them as menacingly as I could. 'I can't be responsible for any injuries if my dog attacks.'

Sesi had grown tired of circling the two youths. Worryingly, they both seemed quite comfortable with her being so close. When the short one extended his palm for her to lick, she trotted over obediently, and sat beside him. The young man looked triumphant. He eased into a grin, but it had a vicious undertone. Then he gestured at Roxi.

'Did you say that's a minipig?' he asked. 'I thought they were supposed to be tiny?'

'If it's for real then we can sell it,' his accomplice added. 'Do we have space in the back of the van?'

'You can't take her!' I said, and took a step forward. 'She's special.'

'She's smashed,' observed the larger youth. 'It's what happens when pigs eat fermented apples. I learned that at agricultural college. Before they kicked me out.'

'Just go.' I raised my voice once more; affronted by their apparent lack of concern at being caught red-handed. 'Get out of here!'

'*Shut it,* Papa Smurf!' The little guy really flared up here. His eyes flashed and he raised his hand as if threatening to slap me from there. I looked to Sesi for support, but she seemed confused. 'Your minipig is coming with us.'

A tussle of any kind was the last thing I was cut out for, but just then I knew that I would do whatever it took to protect Roxi. I squared up to the short lad, aware that his more intimidating partner in crime was looming behind him, only for his attention to switch to something towards the far side of the patio. When Sesi barked excitedly, I turned to look around.

'Nobody calls my friend a Smurf,' said Tom, quite calmly, while weighing a monkey wrench in his hands. 'He's gone to a lot of effort to look like a Na'avi warrior. Show some respect.'

The short guy turned to his accomplice. All of a sudden, in the face of Tom's arrival, his cocky demeanour just shrank away.

'It's Jake's old man,' he muttered under his breath, and took a step backwards. 'We should go.'

Tom frowned for a moment, and then beamed at them in recognition.

'Well, *hello* boys!' he declared, commanding their attention. 'You used to give my son a hard time at youth club, right?'

'Not us.' The bigger of the two shook his head so furiously that his jowls took a moment to catch up. 'You got the wrong guys.'

'Oh, I don't think so,' said Tom, beaming still like this was some kind of happy reunion. 'Did you know that Jake joined the army? I'm guessing that you didn't do anything so constructive with your lives, right?'

By now, the shorter guy appeared to be distinctly uncomfortable inside his own skin. He was moving about like someone torn between fight and flight. With his big friend looking visibly subdued, I guess he realised the former was not an option when faced with one man and his monkey wrench.

'This is crazy,' he said. 'We're leaving.'

'*Not just yet*!' Tom's instruction served to halt them both. When they faced him once more, all the sunshine in his manner had vanished. 'I believe we all know that's my Rosie in the back of your van,' he said. 'I want her back right now, and if I find you've touched her in any way, mark my words this wrench will touch you back.'

By now, even the larger boy was beginning to look stressed and worried. They might not have known that Tom had a pet name for his stone-cutter, but plainly they realised who they were dealing with here. Only Roxi seemed oblivious to what was going on. Her hearing was just fine, but her sense of smell was the motor that powered her. She had returned to her excavation. There, she was busy attacking the soil around the pipe with her snout. Much of it she flung over her head. It meant sprays of dirt kept raining over the patio. When a clod hit Tom on the cheek, however, it caught him by surprise. 'Easy, girl!' he said, glancing at the minipig.

It was all too much for the two young thieves. With Tom momentarily distracted, they made a break for it by running wide around Tom for the drive way. Caught up in the excitement, Sesi gave chase this time, but only for a couple of yards before deciding that Tom was more fun to be with.

'What now?' I asked as the pair disappeared on us.

'Call the police,' said Tom, before taking off after them himself.

'Wait!' I called out. 'I can't do that!'

Tom slowed and turned to face me. 'Why not?'

I asked him to look around. 'Roxi has caused criminal damage here! Criminal damage without a movement licence. Those kids haven't actually taken anything, but I'm worried we could lose our minipigs if this went any further.'

'She won't go down for this,' he said, spreading his hands, 'although I accept that Roddie might take a very dim view.'

'Please,' I said. 'Can't we just deal with this ourselves?'

Tom looked over his shoulder. 'Well, I've blocked the drive. If it makes you feel any better, they're going nowhere for now.' Swinging the monkey wrench from one shoulder to the other, he prepared to set off around the house. 'You make sure Butch and Roxi are safe and sound, I'll see to it these boys don't trouble us again.'

'Tom!' I stopped him in his tracks. 'I don't think we should be adding assault and battery to the crime sheet here?'

He seemed a little surprised by my note of caution.

'I'm not going to *hurt* them,' he said, asserting his grip on the wrench. 'I'm all for giving people second chances.'

As we spoke, Roxi clambered from her hole and stood groggily before us. I knew she viewed me as a human feeding machine, but this was no time for treats, alcoholic or otherwise.

'Looks like our luck has run out,' I muttered. 'Roddie has put up with the noise from our minipigs, but this will push him over the edge.'

'Just tell him the truth,' said Tom. 'Roxi foiled a robbery.'

'And trashed his garden in the process.' I shook my head. 'Trust me, Tom. In my neighbour's eyes, Roxi will come out of this as bad as any burglars. For another thing, it'll freak him out. He's *terrified* of being robbed.'

Tom sighed to himself. He looked around at the scene of the crime.

'Then call Emma,' he said. 'Ask her to keep Roddie occupied for half an hour. We can make this good again. As landscaping goes, it isn't as bad as it looks.'

'You'd do that?'

'With *proper* help this time,' Tom warned. 'No standing around or sloping off to make the tea. We all need to break sweat to bring this back up to speed, but it's for the best. I think perhaps you're right about your neighbour. It wouldn't do him any good knowing that intruders have been here, minipig or otherwise.'

The sound of the van's engine turning over reached us from the drive just then. Whoever was behind the wheel gunned the throttle, as if threatening to ram Tom's Land Rover out of the way. I very much doubted they would see it through. I just hoped my friend would be true to his word when it came to dealing with them. I looked down at Roxi. She hiccupped, so forcefully it caused her to stagger sideways, but she'd calmed down nonetheless.

'You need to sleep it off like Butch,' I said, and set about dialling Emma's mobile phone. 'Somehow I need to explain what's happened here.'

It took some effort to coax Roxi back to the pigsty. I managed to lure her there with the help of a lint-coated mint that I found in the corner of my pocket. For a change, Sesi's shepherding instincts served to help me in the task. Even so, every couple of steps, something else would draw Roxi's attention and she'd stagger off at a right angle.

By the time we made it to the orchard, I turned and noted

Tom was sitting between the two young men inside their van. I could hear him holding forth. They seemed cowed by his presence, but were fixed on every word he said. The chickens, meantime, just looked appalled at the shambolic state of the minipigs. Butch still hadn't stirred. It looked like only the hangover would wipe the smile from his whiskery face. With Roxi safely over the fallen fence, I decided it was time to rouse him. It took a few attempts. I had to call his name repeatedly, slapping him lightly around the snout. Eventually, with a groan, Butch opened his eyes. He seemed shocked to find me looming over him, and scrambled to his feet. It didn't last long. Having taken himself straight to bed inside the shed, I heard him crashing onto the straw. Roxi followed suit, messing up only momentarily by hitting her head on the top of the opening. The chickens followed, fanning wide and checking behind them at every opportunity. When the last one joined the plastered pair, I turned to assess what had to be done.

'This is bad,' I muttered under my breath, 'but I suppose it could've been worse.'

I felt shaken by what had happened, but anxious to keep busy and repair the damage. As I struggled to put the fencing back together, I was surprised to hear Tom's Land Rover start up and pull away. At first, I thought perhaps he was moving it to let the van go. Instead, his was the only vehicle to accelerate down the lane.

I was underneath the fence at the time, supporting it with my head and hands. All of a sudden, I felt a little bit alone. More so when I heard a rustle of leaves from the orchard on the other side. Without doubt, the two young men I had confronted were approaching. I was trapped. I couldn't just drop the fence and flee without risk of being flattened. Nor did I have the strength to push it with any force so it flopped on top of them. Why had Tom abandoned me, I thought in a panic? Was this some stupid test?

'OK, mate. We can take it from here.'

Before I could work out what was going on, two pairs of hands appeared at each side of the fencing. Then the pressure on my head lifted away. I looked around. The big lad and his sidekick were visible in the triangular gaps between the panel and the posts. Sesi visited each one in turn. I just looked lost.

'Tom will be back in a couple of minutes,' said the smaller youth. 'He's just gone to pick up some rolls of turf from his smallholding. Providing we help you out here, he's prepared to turn a blind eye to everything that's happened.'

'Really?' I turned to his accomplice. Despite his size he looked like he was set to shrink at any moment. 'You're lucky,' I added. 'Tom is a generous man.'

'And I'm sorry I called you Papa Smurf,' he admitted, and bowed his head. 'It was hurtful.'

I studied him closely. 'Tom made you say that, didn't he?'

He nodded, without making eye contact, and then helped his friend to position the panel. I smiled to myself as they disappeared from view, and then held the fence so they could hammer in nails and make everything safe again.

Surprisingly, we made a good team. Once Tom had returned, I grafted as hard as the two youths to restore Roddie's lawn to its former glory. They didn't want to talk, or even offer their names, but to be fair both boys worked to pay off their debt to Tom. I just put my back into the task because I knew that if Roddie found out Roxi had escaped then things would change for the occupants of the pigsty. And not in a good way.

By the time we had finished, blue face paint was trickling from my brow into my eyes.

'Thanks,' I said, and extended my hand to the two lads. They nodded in response, only to look pleadingly at Tom.

'Job done,' he confirmed, having worked closely on making the replacement turf indistinguishable from the old. 'I'm always looking for a capable pair of hands to help me out. You've shown you're prepared to work, and evidently you know your way around landscaping equipment. Now, I'm snowed under with pending jobs at the moment. Next time, if you're interested, I might be able to offer cash in hand.

'Really?' I cut in. Tom was addressing the two young men, but frankly I was stunned. Not least because he'd never offered *me* the chance to work outside every now and then. 'You'd do that?'

'Why not?' Tom clapped the smaller youth on the shoulder. 'Off you go, lads. It's high time we all moved on.'

Together, we made our way to the mouth of the drive. The boys set off at a lick in their van, while Tom invited me to jump into the Land Rover and transported me the short stretch to our house.

'Thanks,' I said, as he pulled in. 'I think we've spared an old boy a whole lot of grief.'

'It's our secret,' he said, lifting the handbrake. 'What did you tell Emma?'

'I told her the truth,' I said.

'Including the fact that Roxi isn't pregnant?'

Tom asked this in a way that told me he knew the answer already.

'I left that out,' I confirmed. 'Along with the fact that they're both asleep in a stupor.'

'Then you'd better hope they don't wake up for a good while. Otherwise, she'll see that for herself,' Tom pointed out. 'And given that she believes Roxi is pregnant, that's going to concern her unnecessarily.'

'I know,' I said dolefully. 'It's now or never, isn't it?'

Tom didn't answer. There was no need. I climbed out of the Land Rover, thanking him once again, and prepared to make an honest man of myself.

I was in a mess with minipigs. We didn't own two pint-sized porcine pets that snuggled up to us on the sofa. Quite frankly, what we had here were a pair of flatulent inebriates who were sleeping off their binge at the back of a converted garden shed.

As for me, in my metaphorical hole, I realised I had reached rock bottom. In a bid to meet my wife's insatiable desire for a big family, beyond children and then animals, I had offered Emma snake oil and she had swallowed it all. Like me, she had learned from experience that minipigs didn't stay so mini. It was just something we never mentioned as I sought to create more. If I understood my wife, what mattered to her was the prospect of new arrivals. Lou hadn't helped the situation. She'd come home from school recently with a list of friends who had put their name down for one of Roxi's offspring.

Worse still, I couldn't deny to myself that I had been drawn by the opportunity to make easy money. Even though the end result could've saddled my buyers with pigs that kept on getting larger, what had mattered to me was the prospect of a flash holiday. In many ways I wasn't much better than the two reprobates I had caught in Roddie's shed. Tom had given them an opportunity to make amends and even prove their worth. It was an approach that truly impressed me. I just hoped that my wife would show the same degree of humanity by giving me another chance.

'Can't it wait?' This was Emma's first response when I asked if we could speak. She had rushed in looking flustered and anxious. All four kids had caught the sun, which apparently was my fault for asking her to keep Roddie occupied. 'I've been so worried,' she said. 'Poor Roxi. *Anything* could've happened!'

'What matters is she's safe,' I said, and steeled myself to bring her up to speed. She had found me at the kitchen table. I was sitting in front of my shaving mirror with a bottle of baby lotion on one side and a heap of blue-stained cotton pads on the other. As Lou and May headed upstairs, Emma ran herself a glass of cold water. Sitting down opposite me, she looked directly into my eyes.

'You owe me big time for keeping Roddie in conversation. He's exhausted me! I now know all the planning applications going on in this village, plus every last scrap of gossip. Much of it, I should say, revolves around you.'

'Emma, there's something I want you to hear from me directly.'

'I know it's inevitable, seeing that you're at home all day,' she continued, talking over my bid to steer her onto a more sensitive subject. 'Even so, you and I need a chat about looking respectable when you open the front door to people. From the accounts he shared with me, people are beginning to suggest your mental health is deteriorating.'

I finished wiping the last trace of gold from my eyelids. My face was still tinged with blue. I looked like someone who was about to be sick, but that wasn't entirely down to the face paint residue.

'I'm fine,' I told her. 'I'm just a little stressed right now.'

'*You're* stressed? How do you think I've been feeling? You call with talk of missing minipigs and garden thieves . . . why didn't you just phone the police?'

'Tom dealt with it in his own way. We didn't want to worry Roddie.'

'But it's fine to worry me?' Emma set the glass down and rose to her feet. 'I need to see Butch and Roxi,' she said. 'This afternoon has made some things very clear to me.'

As she headed for the back door, I looked long and hard at my reflection. Steeling myself, I then set off to catch up with her.

The little ones were playing on their bikes in the yard. By the time I had picked myself out of the bush, having been forced to take action to avoid a collision, Emma was already through the gate. I found her in the pigsty. She was crouching in front of the entrance to their shelter, frowning.

'It's been an ordeal for us all,' I said. 'They're just dozing.'

'But they *always* come out to say hello. It doesn't matter how tired they are.'

I cleared my throat.

'OK, they're not just tired,' I confessed. 'They're drunk.'

Emma looked up at me, then slowly climbed to her feet. She laughed dismissively, but that didn't last long.

'Tell me that isn't true?'

I felt a breeze pass over me just then, which was weird because the day had been so bright and still. Patiently, I explained about the fermenting apples Butch and Roxi had found on the other side of the fence. I also knew that their intoxicated state wasn't really the issue here. As far as Emma was concerned, Roxi was in pig.

'You really don't have to be concerned about the effects,' I said to finish, and prepared to break the news.

'Of course we should be concerned!' Emma looked outraged. 'We've had four children. Each time, I've given up drinking because of the risks to the unborn child. That's thirty-six months without a *single* drop of alcohol so that we could have healthy children.'

'That's not entirely true,' I pointed out, before I could think and then stop myself. 'What about the day before you found out you were pregnant with May?'

'What about it?'

'You went to that Primal Scream gig with a couple of friends—'

'That was a one off,' she cut in. 'And for obvious reasons, I don't remember the details.'

I drew breath to remind her, but thought better of it. The band had played two summer nights, in a circus tent in a park near where we lived in East London. With every babysitter booked up in the area, and Lou just nine months old, Emma and I had decided to cover for each other. I had gone on the first night, with the husbands and partners of Emma's friends. The event had been sponsored by a brewery company famed for their strong lager. As a result, the drink was subsidised, which went some way to explaining why I had returned home so unforgivably slaughtered that I slept on the kitchen floor. The next day, Emma had lectured me about the fact that I was no longer a teenager. Why couldn't I just appreciate good music without a beer in my hand, she had asked, before setting off for her night out.

Several hours later, the front door had crashed open. I leapt from the sofa to investigate, and found Emma crawling over the doorstep. Something had happened to the world, she insisted, as I helped her upstairs. It had slipped on its axis, she claimed, before passing out on the bed. I would've gladly discussed this conviction of hers the next day. What stopped me was the fact that when Emma surfaced she came downstairs with a blue-banded pregnancy testing stick in one hand

and a grim expression across her face. Her doctor had assured us that it would be fine, but for nine months we were worried.

Facing Emma now, I could see the same faintly haunted look she had carried with her throughout that time.

'It didn't do May any harm,' is all I chose to say. 'And believe me it won't be an issue for Roxi.'

My assurance did nothing to change Emma's expression. If anything, much to my surprise it prompted her eyes to moisten.

'It's not just that,' she said. 'When you called to say Roxi was missing, I realised just what these minipigs have done to us. They're turned our lives inside out! Every day brings another challenge, stress or strain. Now we've had a full-blown crisis and it's really brought some truths home to me. Matt, we can't go on like this any more.'

I took a step back, feeling a rising sense of shock.

'Are you trying to say,' I asked quietly, 'that you're leaving me?'

Emma laughed, out of nowhere, it seemed to me. It caused a tear to slide down her cheek. Crying didn't come easily for her, but when it did it came from the heart.

'What I'm saying,' she continued,' is that Butch and Roxi have made our lives *complete*.'

I blinked in response. All other brain function was devoted to working out what this meant.

'Go on,' I said after a moment.

'If I'm honest with you, while I was keeping Roddie occupied I couldn't help wondering whether we were doing the right thing by trying to bring more minipigs into the world.'

'You were?' I was still stunned by the brief moment when I thought Emma was trying to call time on our marriage. Now I had to deal with the fact that she was having second thoughts about breeding Roxi. It was the last thing I expected her to say, and the only thing I wanted to hear. Just then, I realised I was grinning, which wasn't appropriate seeing that Emma had started to sob.

'This is all my fault,' she said, as I took her in my arms. 'I don't want to spoil what we have.'

'Trust me,' I said, barely able to believe my change in fortune. 'Nothing is going to be spoiled here. Everything will be just fine.'

'But that's not true.' Emma pulled back to face me. 'Already I'm worried about what this pregnancy might mean. Not just for Roxi, but for us. These minipigs have truly tested our family, and we haven't even been through a winter with them yet. I recognise it's made us appreciate what we have here. But after everything we've been though, I'm just worried that any more could be our undoing.'

My wife was genuinely troubled. It was a rare thing; a first, in fact, and I didn't like it one bit. Maybe I wasn't alone in picking up on this, because first the chickens hopped out to see us, followed by Butch and then Roxi. Butch was wearing what looked like a toupee made from straw. He looked as bleary as Roxi, who sniffed at the air with her snout, before proceeding to wee quite magnificently all over the concrete. Without further delay, I faced Emma once more, and took her hands in mine.

'Roxi isn't pregnant,' I told her. 'Once she's over her hang-over, the only thing she'll be expecting is food.'

Emma looked like she was reeling on the inside. She switched her gaze to my mouth and then back to my eyes, as if seeking some explanation.

So, I told her. I went right back to the moment I had messed up with the second straw of semen, and how a crisis of conscience had persuaded me to jettison the last one. It took a while to account for my reasoning. I didn't need to tell her that these minipigs would continue to test all manner of boundaries beyond just the fence for many years to come. What mattered was the fact that they had brought us to this moment.

Throughout, my wife listened and then laughed, and I laughed along with her. In that time, the little ones decided to wheel their bikes onto the lawn. Honey was first to cycle full circle around the rabbit runs. Frank was next. He steered into the sides a couple of times. The rabbits didn't like it, but they'd survive. Summoned by the noise, the dog arrived at the gate and began barking in a bid to join in. She made quite a noise, but had to compete with Lou, who had chosen this moment to test the volume on her bedroom stereo. To my ear, the noise made by Sesi was more agreeable than Lou's brand of boy band ballad. Still, she was enjoying herself, and there would be other times for me to sing along until she turned it off. Miso was nowhere to be seen. If May hadn't gone out then he was most likely to be in her lap. Since coming off the tranquillisers he'd certainly returned to us in spirit. Naturally, our cat still regarded everyone with the same disdain as ever. His only exception was the one family member who had looked out for him throughout his time in the drug-induced wilderness. As for Butch and Roxi, they milled around us while we talked. They were unusually subdued, but we both knew that wouldn't last. Once they'd had a good night's sleep, I could expect my usual wake up call.

'You know what?' said Emma, as we left them rooting in the pigsty. 'I've come to quite like the early start. It means I make the most of each day.'

'Really? That squealing never feels like the break of dawn to me, more like the beginning of the End Times.'

I closed the gate to the pigsty behind me. Turning to follow my wife back to the house, I found her waiting for me.

'Just a thought,' she said. 'Would it help your mornings if we got a rooster?'

I chose not to answer, stepping aside instead as Honey and then Frank zipped between us on their bikes. Already the circles they were turning had gouged tyre ruts through the grass, but I was beyond caring about that any more.

They could ride until bedtime, so long as everyone was happy.

An Epilogue to the Pigs

I felt nothing but relief following that afternoon. It was like I had returned to being my old self again, but with my spirit renewed by what we had been through. I still had to break the news to the kids, of course, but at least I had Emma onside. The little ones were upset, but it was nothing two ice lollies from the freezer couldn't solve. As far as May was concerned, it simply reduced the chances of Miso relapsing. Only Lou took some time to come to terms with my decision.

When I told my eldest daughter she should bin her waiting list for minipiglets, she didn't speak to me for three whole days.

Thanks to our efforts next door, Roddie knew nothing about how close he had come to losing his mower, gardening tools and even his lawn. Whenever the sun shone, I would hear him giving the grass a little trim, which I took as a sign that all was well. He even called round at the end of that week. I was in the office at the time, and heard the front gate swing open. It meant by the time I opened the door I was well prepared to greet him.

'Well, hello!' I said in greeting, having purposely thrown on Emma's apron with the body of burlesque dancer on the front. I rested a hand on my hip, smiling genially. Roddie paid me no attention whatsoever. Instead, I found him looking with some disdain at the little pile of rabbit intestines at his feet. 'That'll be the work of our cat,' I told him. 'Miso's not been himself for a while, but he's in fine health now.'

'I must say,' he replied, without looking up, 'I'm not fond of wild rabbits. They have a habit of digging up my lawns.'

'*Anyway*,' I said, moving on quickly, 'how are you?'

Finally, Roddie faced me. His eyes dropped to my apron for a moment. I hoped that meant word would soon spread across the village.

'I am very well,' he said, as if to suggest by contrast that I wasn't. My neighbour had a wicker basket with him. It was filled with apples from his orchard. They were even more rotten than the ones that Butch and Roxi had binged on. 'I was wondering whether your minipigs would like these,' he said, and offered me the basket. 'They're no use to me, but I'm sure they would enjoy them.'

'I'm sure they would,' I said, as he passed me the basket. 'In moderation.'

As we talked about the success of the fete, and the under-performance of the face-painting stall, a Land Rover swept over the crest of the lane and sounded its horn. I looked up as it passed to see Tom waving from the driver's seat, with two familiar faces sitting alongside him. Even Roddie waved back.

'He's a decent man,' he said, as if perhaps this was someone I should aspire to become. 'It's high time I bought another tasty half pig from him.'

'That's probably the easiest way to keep them,' I told him. 'Whole pigs can be a handful.'

Roddie turned back to face me. I'm sure he heard my comment. He just didn't respond to it. Instead, something in the house behind me drew his attention. I knew what he was looking at. I didn't feel as if I had to justify why two minipigs were peering cautiously from the front room. In a bid to get across how relaxed I was with the situation, I just smiled at him.

'I should be getting on,' he said, eventually. 'You can keep the basket until you're finished with the apples.'

'Thank you,' I said, 'on behalf of Butch and Roxi. It's a very kind gesture.'

Roddie glanced at the minipigs once more, as if to be sure of what he had found here.

'They really are most peculiar creatures,' he observed. 'I still hear them over the fence, you know.'

'Sorry,' I said automatically, and focused on the rabbit entrails.

'There's no need.' Roddie waited for me to look up at him. It was a rare thing to see him smiling so freely. Just a trace, but as good as anything I could expect. 'I can see they make good company.'

As soon as he'd left, I headed for the front room. There, Butch had gone back to making himself comfortable on the sofa, while Roxi was already stretched out on the floor, all but matching it for length. I showed them both the basket.

'Now be good,' I said. 'So long as nothing gets damaged, chewed or eaten while I'm working, you can each take an apple back to the pigsty. This is about trust, right? Don't let me down.'

The minipigs had settled in a part of the house where the sunshine was flooding in. I had learned very quickly that this was a good thing. The light and warmth served to subdue them nicely. Even so, that wasn't solely responsible for keeping them out of mischief. Just then, Roxi picked up on the smell of the apples. I just told her to stay put. I could be fairly sure she and Butch would do as I asked. In fact, I had enough confidence in them to go back to work in my office for a while. As it had turned out, they really were smart animals, just as Emma had said. In this case, following the incident next door, and a controlled test with some more old apples, we had discovered that we owned two minipigs who would do anything for a tipple.

Emma was the reason that Butch and Roxi had been allowed back in the house more frequently. Her feeling was that we could've lost them for good when the fence fell down. This was her way of letting them know that they would always be welcome here.

Privately, I thought it was a loopy idea. As far as I was concerned, it meant the minipigs were closer to the contents of the kitchen, and my fast track to a criminal record. Much to my surprise, however, Butch and Roxi came into their own as house guests. They were never obliged to come inside. Every now and then we simply opened up the pigsty and the French windows, but almost always they made a beeline for the front room, if only for the biscuit crumbs.

Whenever real visitors dropped round, to be greeted by friendly grunts, they left with no idea of what we'd been through. My brother and his young family were enchanted, as was my sister's daughter when she laid eyes on them. Emma's mum, who had found herself in later life by becoming a loyal and loving grandmother, took to phoning on a regular basis so the kids could update her on Butch and Roxi's exploits. My dad just made bacon jokes, and then spent quite a length of time admiring Tom's handiwork with the shed.

Our rules for allowing the minipigs inside became more refined with experience. They only joined us during dry weather, and after a meal, when hunger wouldn't lead them to try to dig through carpets in search of a thirty-year-old crisp shard caught in the floorboards. My biggest worry was that one of them would flood the place in urine. Frankly, they had massive bladders, and I really didn't fancy mopping up after them. Emma asked me to give them a chance. Maybe they had matured, because the floors remained unspoiled. I did wonder whether Butch and Roxi just held on because they enjoyed being in the house. Especially if there was a sunny spot on offer. Either way, with no more reason to fill a bucket of soapy water, it was nice to have them around.

At first, I really didn't like leaving them entirely unattended. We'd only suffered one incident when they were left to their own devices. That resulted in the disappearance of my sole surviving videogames controller. Emma and the kids swore they hadn't seen it. As the controller had gone missing while the minipigs were indoors, the finger of blame turned to Butch. Despite a search of every room, as well as the pigsty, it failed to show up. I had prepared to be livid with him, but decided instead that perhaps it was time I retired from what was essentially a young man's pastime. Increasingly, I found the only hours available to play came when everyone else went to bed. Frankly, after a day balancing work, children and animals, I was usually snoring on the sofa before Emma turned in herself.

What I liked best were the times Butch and Roxi joined us when everyone was in the house. It was, of course, utter bedlam. Alone with me, both minipigs were relatively calm. Throw children into the mix, and it wasn't just the little ones who became excitable. Surprisingly enough, it was Lou who hyped them up. She was always whipping out her camera in order to update all social media of their presence in our midst. Her exploits gained quite a following. Not because she had pictures and clips to display of two tiny minipigs in teacups.

She didn't have anything like that, in fact. What drew her audience was the fact that two farm animals were welcome in our house, one of which was almost as big as the furniture.

As for the sofa, I no longer had a space I could call my own. Roxi was barred from it, on the grounds that she might break the springs. Instead, splayed out on the carpet in front, she served as a footstool for Emma and the kids so long as they were gentle. Butch enjoyed a seat, particularly if it meant he had a lap beside him to rest his head. Me? I retreated to an old rocking chair that sat in the far corner, one that nobody ever used for anything but dumping newspapers.

With the supplements serving as a cushion, I found it wasn't that uncomfortable. More importantly, it afforded me a little space, peace and quiet, as well as a clear view of my family.

Postscript

Shortly after I finished writing this book, Miso was struck by a vehicle on the lane. I was driving home with the little ones just after it happened. The car responsible had stopped a few metres beyond the writing animal, only to pull away as I rushed out to help.

Our cat died in my arms, with Frank and Honey looking on. There was nothing I could do for any of them.

I was determined to see him rest in peace. Butch and Roxi often roamed the garden, which meant no plot was safe, and so I called upon Tom for help. It was a measure of his kindness that he offered to take care of things, and invited me to meet him at his smallholding. Leaving Emma at home to comfort four distraught children, and with Miso's body under a picnic blanket in the boot of the car, I set off down the lane.

I was expecting to find Tom waiting for me with a spade in hand. Instead, as I drove through the gates, I came face to face with the man astride his mechanical digger.

'Isn't this a bit extreme?' I asked over the chugging motor.

'Not if you want to get the job done properly,' he said, before swinging the cab around to face the other way and instructing me to follow him.

* * *

That evening, having carefully softened my accout of the burial ceremony, I sat around the kitchen table with my family. In a bid to make things easier for the kids, Emma had allowed Butch and Roxi to join us. The hens didn't seem too happy about being separated so late in the day from the swine they were designated to protect, but we had to assume they trusted us. Despite their presence, and with the dog whining softly behind the bars of the child gate, everyone stayed lost in thought for quite some time. For a cat who didn't much care for our company, Miso's absence from the house was felt by us more than ever before.

Acknowledgements

This is a personal story, but so many people were involved in its creation. Firstly, I should like to thank Gillie Russell and my literary agent, Julian Alexander, for not laughing me out of town when I first discussed the idea. Similarly, I am indebted to my editor, Rupert Lancaster, Kate Miles and the entire team at Hodder, for truly believing that minipigs have legs. I can safely say I have enjoyed writing this book more than any other, and this is purely down to their commitment, creativity, good-humour and enthusiasm.

I should also apologise to Philippa Milnes-Smith. She has represented my work as a YA author for many years. *Oink!* is not quite the teen thriller she had in mind. With luck, she'll turn a blind eye.

On the home front, I am of course most grateful to Tom for his help, friendship and wisdom, and also to his family. I just hope that I have done the man justice here, and not painted him as 'some bloody Billy Bob' as he feared. My thanks also to Sarah and Fran for holding the fort, Jojo for the human touch, and also Tony York whose highly-recommended pig keeping course in Wiltshire answered so many of our questions. His wisdom and advice allowed us to sleep at night – on paper, at least.

Finally, from the heart, I must thank my children, wider

family and the friends who have helped me to write this book.

As for Emma, my beloved wife, I am well aware that she could have ordered me to scrap this story at any stage. I was especially worried when I first let her read an early draft, but her support and trust in me has been astonishing. She really is one of a kind, and I couldn't be without her.

Last but not least – and it isn't every day that a writer needs to do this – I have some livestock to thank. Butch and Roxi have taught me a great deal about myself throughout this time. Mostly it concerns the outer limits of my patience, but for all the trials we wouldn't be without them. Even if they keep on growing, they'll always be much-loved minipigs.

Author's Note

I have changed the names of characters and altered locations as well as certain details, but only when absolutely necessary to protect people's identities.